Computer Vision based Identification and Mosaic of Gramineous Grass Seeds

Xin Pan · Xuanhe Zhao · Weihong Yan ·
Jiangping Liu · Xiaoling Luo · Tana Wuyun

Computer Vision based Identification and Mosaic of Gramineous Grass Seeds

China Agricultural Science & Technology Press Ltd.

Springer

Xin Pan
College of Computer and Information Engineering
Inner Mongolia Agricultural University
Inner Mongolia, China

Weihong Yan
Grassland Research Institute
Chinese Academy of Agricultural Sciences
Inner Mongolia, China

Xiaoling Luo
Inner Mongolia Agricultural University
Inner Mongolia, China

Xuanhe Zhao
Inner Mongolia Agricultural University
Inner Mongolia, China

Jiangping Liu
College of Computer and Information Engineering
Inner Mongolia Agricultural University
Inner Mongolia, China

Tana Wuyun
College of Mechanical and Electrical Engineering
Inner Mongolia Agricultural University
Inner Mongolia, China

ISBN 978-981-16-3500-7 ISBN 978-981-16-3501-4 (eBook)
https://doi.org/10.1007/978-981-16-3501-4

Jointly co-published with China Agricultural Science and Technology Press, Beijing, China
The print edition is not for sale in China (Mainland). Customers from China (Mainland) please order the print book from: China Agricultural Science and Technology Press

This Springer imprint is published by the registered company Springer Nature Singapore Pte Ltd.
The registered company address is: 152 Beach Road, #21-01/04 Gateway East, Singapore 189721, Singapore

Preface

Grassland is the refreshable natural resource for human beings, and the dynamic monitor and digitalized government of grassland are vital for the sustainable strategy. At present, the research on the grassland digitalization is rare. Hence, it is of great importance in theory and reality to realize grass recognition by computer vision, which will make contribution to improve the accuracy of grass auto recognition, and data acquisition.

Computer vision is a multi-disciplinary technology that takes advantage of computer and image acquisition system to imitate the human vision in converting image into digital signal. Different from the remote sensing technology-based digital system of the grassland, computer vision emphasizes on digital images of grassland and forage captured by digital cameras. Hence, the book focused on intrinsic feature extraction, so as to realize auto recognition of forage, and microscope images mosaic. The computer vision-based digital platform of grassland and forage was developed by integrating the images mosaic and some related function modules.

The book is an initial effort in the research of grassland digitization based on computer vision, especially focuses on the identification and mosaic of gramineous grass seeds. It is a general research report of our research team in Inner Mongolia Agricultural University and Grassland Research Institute of Chinese Academy of Agricultural Sciences, which aggregates our collective intelligence and efforts in the research. The book was written by Pan Xin, Gao Xiaojing, Liu Jiangping, Wuyuntana, Ma Yubao, and Yan Weihong. Among them, Yan Weihong and Ma Yubao are from Grassland Research Institute of Chinese Academy of Agricultural Sciences, and the other team members are from Inner Mongolia Agricultural University. The final compilation and editing of the whole book was completed by Pan Xin.

This work is dedicated to Researcher Zhu Xu and Dr. Feng Hao for their great contributions. We are also inspired by their deep love of the grassland, which has become the eternal part of their lives. We owe gratitudes to Researcher Liu Guixiang, the Doctoral supervisor in Grassland Research Institute of Chinese Academy of Agricultural Sciences. Thanks to the help and advices from Professor Ma Shuoshi, Professor Xue Heru, Professor Ding Xuehua, Professor Han Guodong, Professor Wei Zhijun, Professor Liu Zhongling, Professor Luo Xiaoling, and Associate Professor Jiang Xinhua. Millions of the researchers and herders work and live in the magnificent grassland, protecting our beautiful home. We are very proud to make all our efforts to be one of them and does something minor to their great career. In addition, we are also grateful at the contributions of all graduates, Chen Tong, Ning Lina, Zhao Xuanhe, Duan Junjie, Zhang Jianhua, Zhai Lin, Han Fan, Zhao Dandan, Cen Yao, Xiang Yang, Wu Lin, Peng Jingjing, Wang Xinyu, Qiao Xue, Guo Yubo, and Dai Xuemin.

This work is supported by the National Natural Science Foundation of China under Grant No. 61562067 and No. 61962048, Central Public-Interest Scientific Institution Basal Research Fund (No 1610332020020) the Postdoctoral Science Foundation of China No. 20100480370 and No. 201104179, the Regional Natural Science Foundation of Inner Mongolia under Grant No. 2012MS919 and No. 2012MS915, the Basic Scientific Research Foundation for Central Public Research Institutes No. 1610332015007, and the Foundations of Inner Mongolia Agricultural University NDPYTD 210-9, BJ09-43, and JCYJ201201.

The research on the grassland digitization based on computer vision is still at the very beginning stage, where there is a long way to explore in future. More researchers are required to get involved in the work. Limited by the professional knowledge in multi-discipline and scientific vision scope, mistakes are inevitable in the book, and we are grateful to the comments and corrections from the readers.

Inner Mongolia, China Xin Pan

Contents

Chapter 1
Introduction

1.1 Digitization Grassland

Grassland is the refreshable natural resource for human beings and provides the basic materials and life guarantee for the rational and sustainable use of human beings. The unique geographical unit and ecological environment have shaped the diversity of rich species and become one of the most important living environments for human beings. However, the degradation situation of grassland in China is very astonishing because of many factors, such as long-term overgrazing, extensive management, climate variation, etc.. At present, the grassland degradation has caused grass yield decline, dust storms, soil erosion and other ecological environmental problems (Tang et al. 2009). The state has attaches great importance to the maintenance of grassland ecosystem, but the mere traditional measurements for grassland monitoring cannot effectively solve ecological environmental problems such as grassland desertification (Wu 2003). Therefore, the dynamic monitoring and digitalization of grassland industry are the basic ways to rationally protect and utilize grassland resources and ecological environment, and further to realize the national sustainable development strategy.

Digital prataculture is a process of realizing digital and visual expression, design, management and control of the relevant objects and the whole process involved in grass industry by using digital technology. It is an important branch of digital agriculture. Within the basic framework of standards and specifications of digital agriculture, it is necessary to construct a digital modern management technology and the theoretical framework, including technologies of computer, network communication, space information, automation and basic subjects of pratacultural science, geography, ecology, to detect, manage and control the elements of grassland ecosystem (environmental factors, biological factors and economic factors, etc.) and vital process. The digital prataculture can macroscopically monitor and evaluate the ecology and economy of regional grass industry, and provide decision support for the sustainable development of grass industry. At a micro level, digital management and control of grassland production process can be carried out by the expert system and knowledge

X. Pan et al., *Computer Vision based Identification and Mosaic of Gramineous Grass Seeds*, https://doi.org/10.1007/978-981-16-3501-4_1

engineering construction, so as to optimize production input, yield and benefit to the maximum extent (Tang et al. 2009). The digitization of grassland is the realized form of digital prataculture. It is the process to store, transmit process and represent the related grassland information in digitized form, and establish the appropriate digital system according to the data models to monitor and manage grassland. Grassland digitization is an emerging research direction to apply information technology into grassland ecosystem management, which has potential values in both theoretical research and practical application.

At present, researches on grassland digitization mainly focus on three aspects: simulation technology of grassland ecosystem process, dynamic monitoring of grassland ecosystem and expert system and decision support of prataculture management (Tang et al. 2009). Images are important digital resources of grassland, which can be generally divided into two categories according to the acquisition means. One is the macroscopic hyperspectral images and remote sensing images obtained by satellites and spectrometers. The other is commonly visible images obtained by ordinary digital cameras.

Remote sensing images can be obtained by using sensors carried on satellite to detect electromagnetic wave radiation information. Remote sensing information is widely used in the study of grassland dynamic monitoring of on account of the merits in its cyclical, macro and systematic aspects, which can be used to monitor the land surface from a macro perspective (Xiao and Zhang 2008) in a medium and long distance. Most of the investigations are based on remote sensing images at a large scale, such as detecting grassland variations (Lausch et al. 2013; Li et al. 2014), and biomass estimation (Potter 2014). However, the overall application of remote sensing cannot meet the practical requirements, such as the synchronous, quasi-synchronous data acquisition, analysis in real-time monitoring and data processing capacity, and automatic recognition of remote sensing images, etc. (Zhao 2003). In addition, the remote sensing images still have limitations in the following three aspects. First, the acquisition cost. Owing to the images are mainly collected by satellite from a long distance, the image number, popularity can not be compared with ordinary digital images directly collected by digital cameras by the researchers. The interpretability of satellite based remote sensing is still a challenge for common investigators. Secondly, the large scale. Generally, remote sensing images fail to capture grassland information from a microscopic perspective, such as individual plant growth status of herbage, composition of herbage species in community landscape, etc.. A comprehensive grassland digital system requires both remote sensing images obtained by satellite and oridianry images by digital cameras. Finally, redundant information. Remote sensing image has a large amount of data and a large amount of redundant information, restricted in data storage, processing, transmission, display and other links. Therefore, the digitization of grassland merely dependent on remote sensing technology is incomplete and one-sidedness.

Digital camera can capture the herbage from grassland from a micro-scale. As the basis and important part of grassland resources, forage plays an important role in providing grassland animal husbandry, environmental protection and soil conservation (Liu et al. 2009). With the popularity of digital cameras, numerous grass images can be easily captured and transferred to computers. Hence, herbage images are as well crucial components of grassland image database and digital grassland. Some of the images can be used for further identification of grass and forage with the help of computer vision. Computer vision technology is an interdisciplinary technology that utilizes computer and image acquisition equipment to simulate human vision acquisition images and convert them to digital signals, and realizes image transmission, processing and understanding and other visual information processing through computer. Image recognition can be realized for automatic identification and classification by the stages of preprocessing feature extraction and feature matching, which form a whole process of pattern recognition. These images are mainly forage plant and community landscape images. Computer vision technologies such as digital image processing and pattern recognition can be used to accurately judge forage species and community component analysis from micro scale, calculate corresponding grassland biological information, and integrate a complete grassland digital system with macro-scale remote sensing images.

Different from the remote sensing technology of geographic information system, computer vision technology mainly studies the digital image acquired by digital camera, which can obtain the required information from a micro scale. This kind of images can be directly collected by digital camera, and then transmitted to the computer for storage and processing, which reduces the error and improves the efficiency of data collection and analysis in grassland field. With the rapid development of image technology, especially the widespread use of digital cameras, researchers have acquired a large number of grassland herbage pictures, which provides a research foundation for the comprehensive digitalization of grassland herbage.

Forage identification is not a simple task due to great complexity among different varieties, which belongs to plant taxonomy, a highly laborious task, consisting of scientific classification of our planet's flora (Yanikoglu et al. 2014).

Therefore, this book focuses on the study of grassland herbage images to extract the characteristics of herbage image with the help of computer vision technology, so as to realize the functions of automatic classification and recognition of grassland forage, and microscopic images mosaic. On the above basis, a digital platform based on computer vision technology was developed to provide new ideas for automatic data acquisition of grassland digitization.

1.2 Computer Vision Based Research Status of Grassland Digitization

In general, there are relatively few studies on digital grassland and forage using image recognition technology, and the methods are relatively simple. Therefore, in this chapter, the main related researches are categorized into 3 types. The first is about remote sensing images—the focus of the studies at present, the second is about the common grassland images captured by CCD digital camera, and the third is about the application in precise agriculture using computer vision.

1.2.1 Grassland Digitization Based on Remote Sensing

Most researchers focus on the processing and understanding of remote sensing images, taking advantages of remote sensing images and geographic information technology to investigate and dynamic monitor grassland resources and environment, such as estimation of grassland biomass, early warning of grassland diseases and pests, and monitoring of grassland degradation. (Wu 2003; Zhang 2007; Qian and Ma 2009). Among them, the Grassland Research Institute of Chinese Academy of Agricultural Sciences started the work earlier, accumulated a lot of research experience. During the 8th and 9th "Five-Year Plan" period, they have undertaken some important research projects, such as National Key Projects, Ministry of Agriculture and National Natural Foundation Project, such as "Research on the Application of Remote Sensing in the Resources Survey of Inner Mongolia Grassland", "A comprehensive investigation and research of Inner Mongolia grassland pasture on protection forest based on remote sensing", and "the research on dynamic monitoring technology of breeding balance between forage and livestock in grassland northern China", etc.. During the 10th "Five-Year Plan" period, the first information professional network system of Chinese grassland resources based on 3S technology, www.grassland.net.cn, has been established in China, with which the queries of 10 ecological information items in forage natural distribution areas. The above works make a frame for digitized grassland.

1.2.2 Grassland Digitization Based on Common Digital Images

However, few researches on grassland digitization based on ordinary visible images, mainly focus on image-based grassland vegetation coverage. For example, Chi et al. (2007) measured the coverage of grassland vegetation by taking close-up images of grassland vegetation with a camera, and measured the coverage of grassland vegetation through 4 steps: data acquisition, image processing, number of pixels extraction and coverage calculation. Compared with traditional coverage measurement methods, it has the advantages of objectivity, consistency and high efficiency and accuracy.In the studies of the impact of grazing on grassland plant communities (Zhao et al. 2010), Cannon PC1057 digital camera was used to capture the images of sample plots under different grazing intensity. The specific method was to divide each area of 3 m × 100 m into 100 consecutive 1.5 m × 2 m, and take 100 consecutive vertical projection photos of each rectangle from a height of 2.5 m. After cutting the photos along the edge of the quadrat with Photoshop, the images of the sample belt were accurately splicing together. Then, the geographic information system software ArcGIS9 was used to make layers of the patches and plant communities in the photos and finally the patch area, plant composition, biomass, density and other indicators under different strengths of the corresponding quadrat can be measured. All above studies show that the grassland images captured by ordinary digital cameras can be used to analyze grassland data, which is conducive to reducing the cost of manual data collection and improving the data accuracy.

To date, only a few researches have demonstrated the feasibility of forage identification. For example, Wang et al. (2010) classified leguminous forage based on shape features of leaf images. The global shape features include the axis ratio, rectangularity, and 7 invariant moments of leaf images. The roughness of the leaf edges are extracted as the local features. Both global and local features are used as the input to probabilistic neural network (PNN) and back propagation network (BP) for classification, yielding correct classification rates of 85% and 82.4%, respectively, in a database of 560 training samples and 1,400 testing samples comprising 14 species.

1.2.3 Researches on the Related Fields

Despite rare research conducted in grass identification, most relevant research on classification of weeds and cultivated crops in precision agriculture provide us some valuable experiences for forage classification. Precision agriculture has been going on for at least 20 years, aiming at the improvement of crop productivity. In comparison, computer vision technology based on visible images has been widely studied and applied in the field of precision agriculture, such as weed identification, pest control, crop growth control, quality monitoring of agricultural products, vegetation coverage monitoring, etc. (Burks et al. 2000; Lv et al. 2001; Ma et al. 2009; Liu 2010), which has the advantages of low cost, high efficiency and high precision.

Natural weed images are used for weed detection and classification to discriminate crops from the weeds (Ishak et al. 2009; Evert et al. 2009) for a reduction in herbicide use and environmental pollution (Burgos-Artizzu et al. 2010). Gerhards and Christensen (2003) proposed a system of site-specific weed control in sugarbeet, winter wheat and winter barley. Bi-spectral images were taken and converted into binary ones to get the contour of the weeds. Color combined with morphological filtering and line fitting were used to segment row crop plants from weeds, yielding highest and lowest classification rates of 92% and 68% over 12 images, respectively (Onyango and Marchant 2003). Meyer and Neto (2008) improved the vegetation index using Excess Green minus Excess Red, instead of commonly used Excess Green, resulting in a higher precision in normalized difference indices (NDI). Burgos-Artizzu et al. (2010) processed images captured in the fields for the estimation of percentage of weed, crop and soil. Tellaeche et al. (2011) developed an automatic computer vision system for the identification of avena sterilis (a special weed seed growing in cereal crops). Wang et al. (2013) proposed an adaptive thresholding algorithm using OTSU and CANNY operators to segment single leaves from leaf images captured from an online system. Yanikoglu et al. (2014) proposed an automatic plant identification system using a rich variety of shape, texture and color features, yielding 61 and 81% accuracies in classifying the 126 different plant species in the top-1 and top-5 choices.

Liu (2010) developed a prototype of rice leaf disease diagnosis system based on image pattern recognition. The recognition rate of 4 kinds of rice leaf diseases and pests can reach more than 83%. Han et al. (2011) put forward pests classification method based on compression perception, extracted by using pest image training area/ standard product /perimeter and 10 morphological parameters, such as red, green, blue, bright, 4 color characteristics of the training sample space of 12 classes of pests and 110 common pests classification. In four different experimental conditions, the correct classification rate of 92.94%, 98.29%, 78.87% and 61.59%, respectively.

Yu (2007) extracted the features of the regional gray level from both sides of cucumber leaf main veins. They established the regression model, from the point of experiment to determine a cucumber point water shortage area.

Li (2009) used machine vision technology to quantitatively evaluate the appearance quality of tea image extraction, and proposed for the first time the classification method of tea based on different particle distribution states. The images based on granular stacking can reflect the geometric features and macroscopic texture features of tea leaves, and improve the efficiency of tea leaf differentiation. 18 shape features and 15 texture features were selected in the study, and the recognition accuracy was 93.8% higher.

Zhang et al. (2009) developed a portable vegetation coverage photography instrument with digital camera as the main collection equipment based on the actual field conditions, and developed the corresponding automatic calculation software to interpret and estimate vegetation coverage.

Owing to seeds are more stable and unlikely to be easily affected by their surroundings, Granitto et al. (2002) assessed the discriminating power of 57 weed seed species. The 12 feature vectors are composed of six morphological, four color and two textural seed characteristics. Bayes Classifier was testified surprisingly good for the classification performance. In their later research (Granitto et al. 2005), a large scale seed base containing 10,310 images of 236 different weed species were used to testify the effectiveness of the algorithm, the recognition rates still reached 99.3% and 98.2% when color images and black and white seed images are testified, respectively.

Shi and Ji (2009a, b) proposed a seed identification system for leguminous weeds. In their work, a seed database of 5,181 microscope images comprising 808 species was constructed. The feature vector composed of 16 components includes the shape geometry and the inner structures of the seeds and umbilicus. The concrete features include the major axis, surface area, perimeter, position and angle between the center points, and 7 Hu invariant moments. The BP Network and SVM classifier are used for classification, yielding a recognition accuracy of 89.29%.

Similarly, identification of nine Iranian wheat seed varieties were conducted on bulk sample images by Pourreza et al. (2012) by extracting 131 textural features from gray level, GLCM (gray level co-occurrence matrix), GLRM (gray level run-length matrix),LBP(local binary patterns), LSP (local similarity patterns) and LSN (local similarity numbers) matrix. LDA (linear discriminant analysis) classifier was employed for classification using top selected features, yielding an average classification accuracy of 98.15% with top 50 selected features. Hong et al. (2015) compared some simple feature extraction techniques and obtained an average accuracy of 90.45% using Random Forest method for the identification of six different rice seed varieties in Vietnam.

It can be seen from the above comparison that visible images based on digital cameras already have a certain foundation in precision agriculture. Nevertheless, in grassland digitization, the overall research lags behind. Most of the micro-scale data collection and measurement of grassland mainly rely on manual completion, with a low automation and subjective observation results. It is urgent to apply image processing and pattern recognition technology to the research of grass industry, especially the research of grass digitization, so as to effectively use information resources and improve the level and efficiency of grass digitization management.

The grassland digitization research based on computer vision can draw on some successful experiences of precision agriculture application, such as the classification of herbage and weed identification, which are very similar in species and traits. The above works mainly focused on the leaves of the weeds and plants. Although forage identification is similar to weed identification in some ways, they differ from each other at least in two aspects, research purpose and identification methods. The former aims at the determination of grassland species for research purpose and further environment protection, while the latter is to remove the weed for less pesticide usage and more grain productivity. So the grass identification mostly focuses on the closely related species of the same families which are similar in shape and appearance, providing valuable information the grassland researchers and farmers. However, weed identification in precision agriculture does not necessarily concentrate on closely related species in the same family. They can be a very large varieties of species widely distributed in any fields where they affect the grain and crop growth whose visual features differ more obviously than forage identification. Due to the different research purposes, weed identification is mainly used to distinguish between weeds and crop, differences between different weeds, weeds and crops are relatively obvious; the grass identification mainly takes the dominant species on the grassland as the research object, to distinguish the different forage species. Especially the different species in the same family are highly identical, which is sometimes difficult for the experts to distinguish.

Therefore, this book mainly uses computer vision technology to extract information of grassland forage image, so as to obtain relevant expert knowledge information of the forage species, and finally establish digital platform of grassland forage based on image recognition technology to improve the level of grassland digital management.

1.3 Research Content

Computer vision based grassland digitization is only in its early stage, and a lot of work needs to be done in theory and practice. Therefore, the research mainly focuses on the visible images obtained by digital cameras, and focuses on the forage identification based on seed images, image Mosaic and the construction of digital platform for gramineous grass in grassland. Among them, the realization of forage automatic classification is the basic element of forage digitization research. Image Mosaic technique is required in practical cases because the limitations by the visual range of camera lens or microscope, multiple images are fused to obtain the overall visual understanding for further studies on forage traits. Finally, on the basis of previous studies, the relevant research results of grassland forage have been integrated into a platform for grassland scientific research workers and herdsmen, named *Digital Information Platform of Grassland and Forage*. Image mosaic can be realized on the platform, which is the highlight of the platform.

This paper is divided into 8 chapters, and the specific content is arranged as follows.

Chapter 1 is introduction. In this chapter, we introduce the background and significance of grassland digitization research based on computer vision technology. Then, the research status and applications of image recognition technology in digital grass industry and precision agriculture are emphatically compared and elaborated, to clarify the important role of image recognition technology in grass industry science. Finally, the main architecture of the book content is listed.

Chapter 2 introduces the concept of forage identification in our research and the image database for experiments.

Chapter 3 presents the identification method of gramineous grass seeds using Gabor wavelets and Local Preserving Projections for intrinsic manifold features. Chapter 4 presents the identification method of gramineous grass seeds using the deviations of local fractal dimensions based on the self-similarity of the seeds. Chapter 5 presents the identification method of gramineous grass seeds using local similarity pattern (LSP) and linear discriminant analysis (LDA). Chapter 6 presents the identification method of gramineous grass seeds using local similarity pattern (LSP) and GLCM, which is an improved method of Chapter 5.

Chapter 7 presents the mosaic method of microscopic image of gramineous seeds. The whole procedure of image mosaic implementation, including image acquisition, preprocessing, feature extraction and feature fusion are introduced respectively.

Chapter 8 introduces the realization process of digital platform based on computer vision. The main contents include the overall framework of the platform, the key technical problems and solutions of the platform, and the implementation effect of the platform.

References

Burks, T. F., S. A. Shearer, R. S. Gates. 2000. Classification of weed species using color texture features and discriminant analysis[J].Transaction of the ASAE, 43(2): 441–448.

Burgos-Artizzu, X. P., A. Ribeiro, A. Tellaeche, G. Pajares, and C. Fernández-Quintanilla. 2010. Analysis of natural images processing for the extraction of agricultural elements[J]. Image Vision Computer. 28: 138–149.

Chi, H. K., G. S. Guang, Z. Z. Xu, and W. P. Yuan. 2007. Measuring coverage of grassland vegetation using remote sensing over short distances[J]. Acta Prataculture Sinica. 16(2): 105–110. (in Chinese).

Gerhards, R., and S. Christensen. 2003. Real-time weed detection, decision making and patch spraying in maize, sugarbeet, winter wheat and winter barley[J]. Weed Res. 43, 385–392.

Granitto, P. M., H. D. Navone, P. F. Verdes, and H. A. Ceccatto. 2002. Weed seeds identification by machine vision[J]. Computers and Electronics in Agriculture. 33, 91–103.

Granitto, P. M., P. F. Verdes, and H. A. Ceccatto. 2005. Large-scale investigation of weed seed identification by machine vision[J]. Computers and Electronics in Agriculture. 47, 15–24.

Han A. T, X. H. Guo, Z Liao, Z. Q. Chen, and J. Q. Han. 2011. Classification of agricultural pests based on compressed sensing theory[J], Transactions of the Chinese Society of Agricultural Engineering. 27(6): 203–207. (in Chinese).

Hong, P., T. Hai, L. Lan, V. Hoang, V. Hai, and T. Nguyen. 2015. Comparative study on vision based rice seed varieties identification[C], 7th International Conference on Knowledge and Systems Engineering. 377–382.

Ishak, A. J., A. Hussain, and M. Mustafa. 2009, Weed image classification using Gabor wavelet and gradient field distribution[J]. Computers and Electronics in Agriculture, 66, 53–61.

Lausch, A., M. Pause, I. Merbach, S. Zacharias, D. Doktor, and M. Volk, R. Seppelt. 2013. A new multiscale approach for monitoring vegetation using remote sensing-based indicators in laboratory, field, and landscape[J]. Environmental Monitoring and Assessment, 185(2): 1215–1235.

Li, X. L. 2009. Reasearch on nondestructive determination of tea quality based on machine vision and spectroscopy techniques[D]. Hangzhou: Zhejiang Uninversity. (in Chinese).

Li, Q., D. W. Zhou, Y. H. Jin, M. L. Wang, Y. T. Song, and G. D. Li. 2014. Effects of fencing on vegetation and soil restoration in a degraded alkaline grassland in northeast China[J]. J. Arid Land, 6(4):478-487.

Liu, L. B. 2010.Research of diagnostic technologies for rice leaf diseases based on image[D].Beijing: Chinese Academy of Agricultural Sciences.(in Chinese).

Liu L. X., G. M. Li, Y. X. Lu, D. L. Wang, 2009. Current status and prospect in the biotechnology of forage breeding in China[J]. Acta Agrectir Sinica, (3): 389-398. (in Chinese).

Lv C. H., X. G. Chen, W. F. Wu, H. X. Zhao. 2001. The technique of detecting and identifying field plants by machine vision[J]. Natural Science Journal of Jilin University of Technology, 31(3): 90–94. (in Chinese).

Ma Y. P., Y. L. Bai, X. Z. Gao, Y. Lu , W. Lei, and J. Du. 2009. Application and expectation of computer vision technology in agriculture production[J]. Chinese Journal of Agricultural Resources and Regional Planning. 30(4): 21–27. (in Chinese).

Meyer, G., and J. Neto. 2008. Verification of color vegetation indices for automated crop imaging applications[J]. Comput. Electron. Agr. 63, 282–293.

Onyango, C., and J. Marchant. 2003. Segmentation of row crop plants from weeds using colour and morphology[J]. Comput. Electron. Agr. 39, 141–155.

Potter, C. 2014. Monitoring the production of Central California coastal rangelands using satellite remote sensing[J] . Journal of Coastal Conservation, 18(3): 213–220.

Pourreza, A., H. Pourreza, M. Abbaspour-Fard, and H. Sadrnia. 2012. Identification of nine Iranian wheat seed varieties by textual analysis with image processing[J]. Comput. Electron. Agr. 83, 102–108.

Qian J. B., and M. G. Ma. 2009. A review of poisonous weeds detection using remote sensing technology[J]. Remote Sensing Technology and Application. 24(5): 685–690. (in Chinese).
Shi C. J., and G. R. Ji. 2009a. Study of recognition method of leguminous weed seeds image[J]. International Workshop on Intelligent Systems and Applications, Seoul, Korea, 1–4.
Shi, C., and G. Ji. 2009b. Study of recognition method of leguminous weed seeds image[C]. in: Proceedings of International Workshop on Intelligent Systems and Applications, 1–4. (in Chinese).
Tang H. J., X. P. Xin, G. X. Yang, B. H. Zhang, X. Wang, H. B. Zhang, and Y. C. Yan. 2009. Advance and prospects in theories and techniques of modern digital grassland[J]. Chinese Journal of Grassland. 31(7): 1–8. (in Chinese).
Tellaeche, A., G. Pajares, X. P. Burgos-Artizzu, and A. Ribeiro. 2011. A computer vision approach for weed identification through Support Vector Machines[J]. Applied Soft Computing. 11, 908–915.
Van Evert, F K., G. Polder, GWAM. Van Der, C. Kempenaar, and L. Lotz. 2009. Real-time vision-based detection of Rumex obtusifolius in grassland[J]. European Weed Research Society Weed Research, 49,164–174.
Wang, J. X., Q. Feng, Y. T. Wang, and X. Q. Shao. 2010. Study on classification for leguminous forage based on image recognition technology[J]. Acta Agrestia Sinica. 18, 37–41. (in Chinese).
Wang, J., J. He, 451 Y. Han, C. Ouyang , and D. Li. 2013. An adaptive Thresholding algorithm of field leaf image[J]. Comput. Electron. Agr. 96:23–39.
Wu, X. T. 2003. Study on methods and application of grassland desertification monitoring by remote sensing [D]. Beijing: Chinese Academy of Agricultural Sciences. (in Chinese).
Xiao L. X., and Z. X. Zhang. 2008. Processes on the boundary definition of agro-pastor al zone in China[J]. Progress in Geography. 27(2): 104–111. (in Chinese).
Yanikoglu, B. A., E. Aptoula, and C. Tirkaz. 2014. Automatic plant identification from photographs[J]. Machine Vision and Applications. 25:1369-1383.
Yu C. L. 2007. Lossless detecting research of Water quantity in the leaves based on image processing technology [D].JiLin: Ji Lin University. (in Chinese).
Zhang H. B. 2007. Study on variation of grassland vegetation based on multi-sensor remote sensing data [D]. Beijing: Chinese Academy of Agricultural Sciences. (in Chinese).
Zhang W. B., B. J. Lu, and S. Shi. 2009, Determination of vegetation coverage by photograph and automatic calculation[J]. Bulletin of Soil and Water Conservation, 29(4): 39–42. (in Chinese)
Zhao, Y. S. 2003. Principles and methods of remote sensing applicaions and analysis [M]. Beijing: Science Publication House. (in Chinese).
Zhao D. L., Z. L. Liu, G. X. Yang, X. P. Xin, C. Z. Liang, W. Wang, L. X. Wang, and W. P. Yuan. 2010. Grazing impact on distribution pattern of the plant communities and populations in Stipa krylovii steppe [J]. Acat Prataculturae Sinica, 19(3): 6–13. (in Chinese).

Chapter 2
Forage Identification and Experimental Materials

2.1 Definition of Forage Automated Identification

The classical plant classification is mainly based on the external morphology of plants, taking advantages of simple observation tools, the experts analyze and compare plants indoors or in the field for their similarity and variability, so as to distinguish and determine the species (Cui 2010). Forage is the most important element in grassland in providing food for cattle and keeping balance of our ecosystem. To protect the grassland resources more effectively, we must know the grass species and their attributes as the first step. Owing to wide varieties between species and similarities of some closely related species, it is hard to identify the forage directly, even for the experts. The current situation hinders further development and practical application for the forage and grassland. To solve the above issues, it is necessary to develop automatic forage identification, so as to further improve the grassland management. Therefore, machine vision is an essential approach to enhance forage images and extract their distinguishing features, thus the accuracy of forage identification can be guaranteed with less dependence on experts.

Computer vision is a newly developed multi-discipline technology to simulate human vision with computer and sensors, with which the images can be acquired and converted to digital signals, and further visual information processing including image transmission, processing and understanding can be realized. In recent years, human biometrics based on the intrinsic information has attracted wide attention as an important ways to solve the public security issues. Many physiological and behavior characteristics of the human body, such as human face, fingerprint, iris, palm print and etc., can satisfy the requirements such as universality, uniqueness, permanence, collectability, acceptability, safety (Zhang 2000; Zhang et al. 2004). Therefore, face recognition, fingerprint recognition and other biometric technologies and products have become the research focus of the industry due to their accurate, rapid and stable identification results (Ho and Chellappa 2013).

Analogy to biometric features used for personal identification, such as face (Wang et al. 2014), fingerprints, palm-prints (Cui 2014) or gait, forage classification can be

X. Pan et al., *Computer Vision based Identification and Mosaic of Gramineous Grass Seeds*, https://doi.org/10.1007/978-981-16-3501-4_2

categorized as a problem of pattern recognition. In fact, each forage species is individually independent and physiologically distinguishable from each other based on the seeds, leaves, and flowers or even the entire plant. As shown in Fig. 2.1, differences in shape, texture and color of forage can be observed in seeds, plants, flowers and other forage organ images. These features can be used as the input of computer vision to achieve automatic classification and recognition. First forage images are acquired by ordinary digital cameras, including seeds, leaves, flowers, organs, and then the image preprocessing, then feature extraction and feature matching process achieve the goal of forage classification by using computer. In the process of research, the image forage database of the living stages in the whole life cycle, from seed to seed can be established. Scanner or digital camera can be used to collect local pictures of seeds, stems, leaves, flowers and other different growth stages to establish the grassland herbage database. By using computer vision technology, the contour, texture, color and other inherent features of the forage image can be extracted to quickly determine the species.

According to the function division of a pattern recognition system, the automatic classification and recognition of forage can be realized by four functional modules, including data acquisition and preprocessing, feature extraction, pattern matching and decision (Fig. 2.2) (Duda et al. 2001; Bian and Zhang 2000).

Data acquisition refers to the use of optical sensor equipment such as digital cameras or video cameras to capture images. The acquired images are disturbed by the external environment, such as uneven illuminations, motion distortion and deformation, so preprocessing the acquired images is necessary in the following. Preprocessing usually consists of two steps, one is to remove noise from the original

Fig. 2.1 Images of forage seeds, plant and flowers

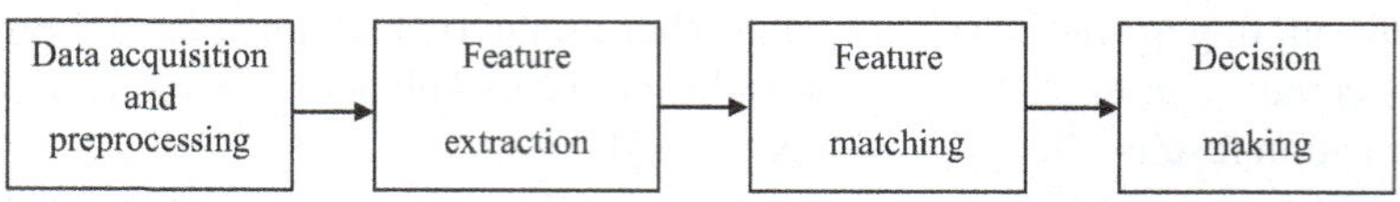

Fig. 2.2 The function modules of forage auto classification system

images with filters and smooth, so as to remove the influence of noise on the images. Then the image is segmented to obtain ROI (Region of Interest) for further processing.

Features are the information that describes the difference between a certain original data and other original data. It is required to satisfy the conditions of the minimum within class distance and the maximum between classes at the same time as possible. Therefore, feature extraction is the most critical part in forage recognition. It determines the recognition performance of the system and becomes the key research issue in the field of pattern recognition. The feature matching module is to match the features to be recognized with the templates in the template database and calculate the similarity (matching distance) between them to determine the recognition result. It is the core problem of pattern recognition and machine learning. The decision module makes a decision on the matching result to determine the category of the tested sample.

In conclusion, forage automatic recognition can be realized by computer vision. Images can be acquired directly by digital camera, which reduces both the difficulties and cost in sample collection and saving. The recognition results with improved accuracy will no longer rely seriously on the judgment of experts, the subjectivity of the judgment results can be solved. In short, forage automatic classification and identification has the characteristics of low cost, high stability and strong anti-noise ability.

2.2 Automatic Identification of Herbage Based on Seed Images

Forage seeds are the material basis for improving degraded grassland, planting artificial grassland and improving grassland animal husbandry production. They are as well as the basic materials for ecological engineering construction, soil erosion and soil conservation engineering construction and urban green space construction in arid and semi-arid areas. Due to the diversity of leaves and flowers, seed traits are relatively stable and less influenced by the environment. Seed morphology is one of the most stable characteristics of plants. Moreover, each forage seed is distinguished from other seeds due to its unique morphological structure (Han and Mao 2011), which is an important basis for automatic classification and identification of herbage seeds based on seeds, authenticity identification of herbage seeds, classification and inspection of herbage.

The external characteristics of forage seeds include shape, size, color, seed coat surface characteristics and their appendages, as well as the position, shape, size, concavity and color of the umbilicus. Traditional classification of forage seeds is mainly based on the external characteristics of seeds with observation, measurement and comparative analysis of seeds carried out indoors by means of naked eye, magnifying glass, anatomic mirror or measuring scale to study their similarity and anisotropy, so as to determine the species of forage seeds (Han and Mao 2011). Our research aims at the identification of individual grass using common CCD images, which provides some valuable information at a relatively small scale as compared to remote sensing images. The research can be categorized to automatic plant identification from photographs, with potential uses in two aspects: one is to identify forage by the images of seeds or other organs for researchers and farmers to protect the forage seeds resources, the other is to assess grass seed to distinguish false seeds and inferior seeds sneaked into seed market from the high quality seeds. The automatic identification system is hopeful to solve these problems.

There are few reports on the studies forage seeds identification. Similar studies include studies on identification of weed seeds and crop seeds, which can provide references for our research. As introduced in Sect. 1.3, the works of Granitto et al. (2002, 2005), Shi and Ji (2009) concentrated on the outer appearance and texture description from the gray level image, such as the color, geometrical features, and etc. Pourreza et al. (2012) chose 131 textural features as the new features for identificaiton, however, the multiple features selected manually for optimum combination are insufficient to represent the seed images of different species. In recent years, some new methods have been applied in seed identification. Hong et al. (2015) used Random Forest for feasible identification results. Cai et al. (2010) used compressed sensing theory to identify 87 species of weed seeds, with the highest recognition rate up to 90.80%.

It can be concluded from the above works that the study on the automatic identification of forage seeds is necessary to be carried out. The shape and texture of seeds are important distinguishing features for seed identification. In addition, the images are presented as pixel intensity matrix of gray scales in spatial domain, whereas the individual pixel based data generally do not provide connection structure and correlation information explicitly. Therefore, more robust global and interior texture features should be explored to improve the classification performance.

Gramineous grass seeds are the essential material basis for artificial grassland construction and grassland improvement. Due to their high yield, good quality, strong palatability and wide adaptability, grass seeds play an important role in China's grassland animal husbandry. Considering the importance of gramineous grass in grassland forage and urban landscaping, we focus on the identification of the gramineous grass seeds. Moreover, despite the small size of grass seeds whose length is about 1 cm, clear images can be acquired by ordinary digital CCD camera, which reduces the collection cost within a feasible collection conditions. Through more subtle observations, it is found the higher similarity between the gramineous seed, especially for the different species in the family, such as Roegneria tunczaninovii (Drob.)Nevski var. macrathera Ohwi, and (b) Roegneria varia Keng shown in Fig. 2.3. Therefore, we emphasized on the textural features in our research, the methods includes Gabor filters, Local Preserving Projections (LPP), fractal dimension, Local Similarity Pattern (LSP), GLCM, and so on. The corresponding contents were involved in Chaps. 3–6. In the remaining of this chapter, the acquisition method and preprocessing of seed images are introduced respectively.

Fig. 2.3 Gramineous seeds **a** Roegneria tunczaninovii(Drob.)Nevski var. macrathera Ohwi, and **b** Roegneria varia Keng

2.3 Experimental Materials

The seeds, 14 species belong to 5 genera (Elymus, Agropyron, Bromus, Roegneria and Stipa), collected from grassland, which are listed in Table 2.1 by name and origin, were provided by the Grassland Research Institute of the Chinese Academy of Agricultural Sciences.

In this section, we will introduce the construction procedure of database for experiments from three aspects, including image acquisition, image preprocessing, and image database.

Table 2.1 The Image database of gramineous grass seeds

No	Scientific name	Origin
1	Elymus nutans Griseb	Ameng, Inner Mongolia, China
2	Elymus cylindricus (Frmch) Hinda	Humeng, Inner Mongolia, China
3	Elymus dahuricus Tunoz	Ximeng, Inner Mongolia, China
4	Agropyron cristatum var. pectiniforme (Roem.et Schult) H Yang	Youyu, Shanxi, China
5	Agropyron cristatum (Linn.) Gaertn	Ximeng, Inner Mongolia, China
6	Agropyron mongolicum Keng	Ximeng, Inner Mongolia, China
7	Agropyron desertorum (Fisch.) Schult	Ximeng, Inner Mongolia, China
8	Bromus inermis Leyss	Ximeng, Inner Mongolia, China
9	Elymas sibiricus Linn	Ximeng, Inner Mongolia, China
10	Roegneria tunczaninovii (Drob.) Nevski var. macrathera Ohwi	Yili, Xinjiang, China
11	Roegneria varia Keng	Milin, Xizang, China
12	Roegneria kamoji Ohwi	WuTai Moutain, Shanxi, China
13	Stipa grandis	Yanchi, Ningxia, China
14	Leymus chinensis (Trii) Tzvel	Changping, Beijing, China

2.3.1 Image Acquisition

The gramineous grass seed images were captured by a commercial CCD camera (DSLR-A350, Sony) under natural daylight illumination inside a room (with non-controlled illumination). The seeds were put on the rough surface of a black mouse mat for less diffuse reflectance, with camera taken photos about 20 cm above. Seeds are generally composed of seed coat, embryo, endosperm and other parts. Gramineous grass seeds are usually caryopsis, in which the embryo is located on the back side of the base of caryopsis (opposite to the side of lemma) with a round or oval concave (Han and Mao 2011). A single kernel gramineous seed has two different sides, the ventral side and the back side. The seed lemma (back side) has relatively uniform and single textures (as shown in Fig. 2.4b) with fewer features. The frontier ventral images (Fig. 2.4a) were used to construct the experimental database because of the multiple inner levels as compared to the back images. Auto flash was used when the light was not strong enough from 5 p.m. to 6 p.m. in winter. The focus lens was less than 100 mm. The camera was set to an automatic focus and exposure. The other options were set automatically under the standard program of the camera. No

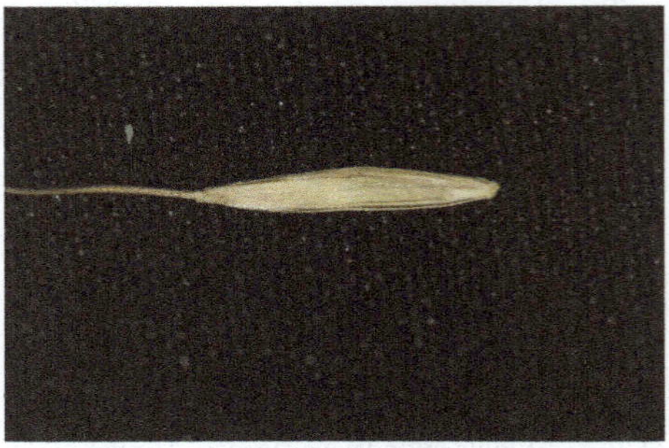

a)

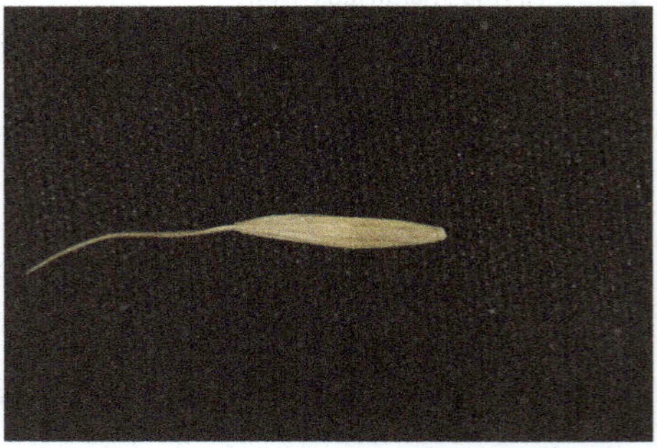

b)

Fig. 2.4 Original seed image (Elymus nutans Griseh). **a** Ventral side, **b** back side

more restrictions in the process of seed image acquisition provide a relatively free style for the users. Figure 2.4 illustrates an original seed image with a pixel resolution of 4592 × 3056.

2.3.2 Image Preprocessing

As can be seen from Fig. 2.4, sparkles and redundant background are present in the image. The sparkles are the stains scattered in the mouse mat which causes noise for further processing. Therefore, the images are preprocessed to eliminate the noise and segment the Region of Interest (ROI).

Preprocessing comprises three main steps. Initially, considering the gramineous seeds are mostly medium brownish without unobvious color difference, the original RGB color images can be converted into gray images for efficient storage and computation. The images were then converted into its binary image with a proper threshold of gray levels (Fig. 2.5).

Subsequently, an open mathematical morphology operation is applied to the binary image to eliminate the bright white sparkles of the background. Suppose A and B represent the image matrix and structuring element matrix, respectively. $A \Theta B$ and $A \oplus B$ denote the erosion and dilation of image A by structuring element B. Then the opening of image A by structuring B can be defined as follows (Gonzalez and Woods 2008)

$$A \cdot B = (A \Theta B) \oplus B \tag{2.1}$$

It can be seen that the opening operation comprises two steps, erosion and dilation. Image A is first eroded by element B to eliminate microscopic spots, and then recovered by dilation. Figure 2.6 shows the result of open operation for a gramineous seed. Lastly, the seed borderline is acquired from the binary image. Rotate the image to keep the principal axis be horizontal, and determine the minimum enclosing rectangle of the seed, thus the ROI is cropped from the original image, as shown in Fig. 2.7.

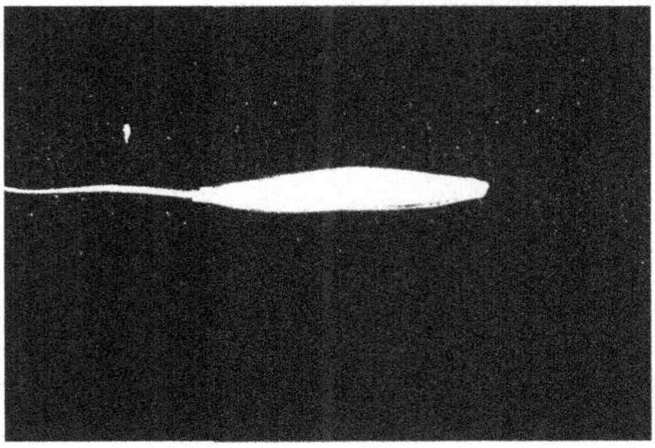

Fig. 2.5 Binary image

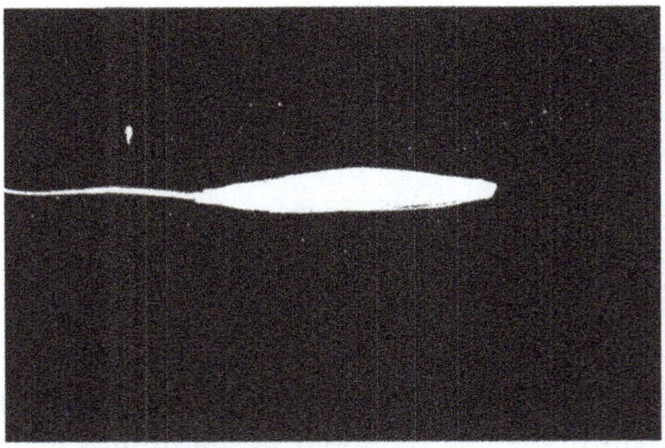

Fig. 2.6 open operation of the seed

Fig. 2.7 ROI after segmentation

2.4 Seed Image DatabaseI

We constructed Seed Image DatabaseI of gramineous grass seed image database using image acquisition techniques described in Sect. 2.2. In view of difficulties in field seed acquisition of the gramineous grass, a database contained 700 seed images comprising 14 species from 5 genera (Elymus, Agropyron, Bromus, Roegneria and Stipa) was constructed.

In the image database, each species was represented by 50 images captured from 10 seeds, 5 images per seed were captured with variations of focuses, orientations and positions. For practical purpose, the classification was confined to the species, rather than each seed, i.e., there were 14 classes in the database, and each class harbored 50 image samples. The original image resolution was 4592 × 3056 pixels of 3.30 MB. The images were stored in a jpeg compressed format. The main bodies of the seeds were cropped without a fixed size due to the seed diversity and the use of the auto focus during image acquisition. Figures 2.8, 2.9 and 2.10 give some examples of the original seed images and their corresponding ROIs in the image database. The ROI resolution and size are listed in parentheses after the figure name. Figure 2.11 shows all the classes in the seed image database of Gramineous Grass.

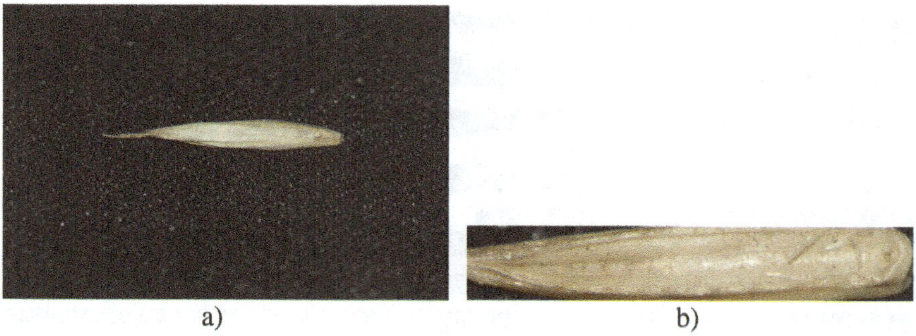

Fig. 2.8 Agropyron cristatum var.pectiniforme(Roem.et Schult) H Yang, **a** original image, **b** extracted ROI(1205 × 170, 30 KB)

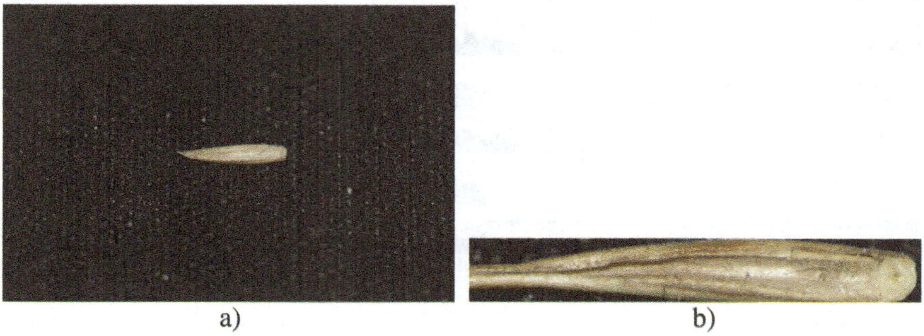

Fig. 2.9 Agropyron cristatum (Linn.) Gaertn, **a** original image, **b** extracted ROI (1002 × 167, 21.2 KB)

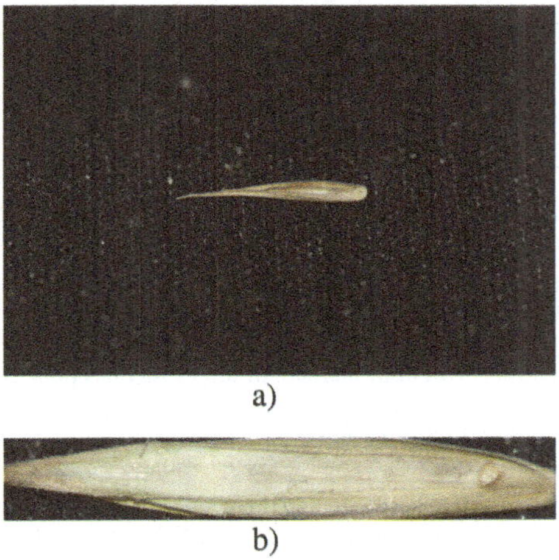

Fig. 2.10 Elymas sibiricus Linn, **a** original image, **b** extracted ROI (1970 × 289, 66.4 KB)

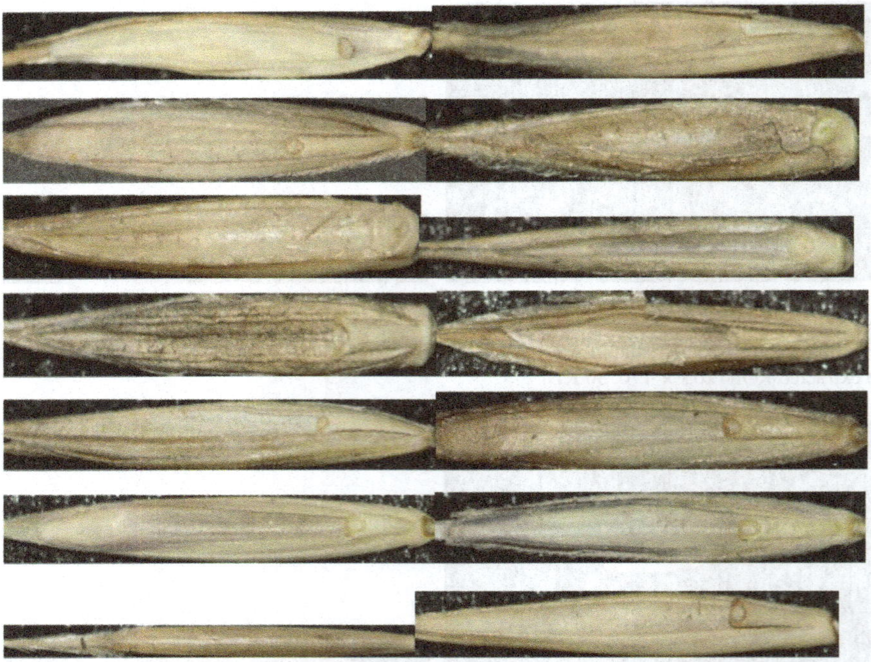

Fig. 2.11 Seed image database of Gramineous Grass (from left to right, from top to bottom. Row 1: Elymus nutans Griseb, Elymus cylindricus (Frmch) Hinda; Row 2: Elymus dahuricus Tunoz, Agropyron cristatum var.pectiniforme (Roem.et Schult) H Yang; Row 3: Agropyron cristatum (Linn.) Gaertn, Agropyron mongolicum Keng; Row 4: Agropyron desertorum (Fisch.) Schult., Bromus inermis Leyss.; Row 5: Elymas sibiricus Linn, Roegneria tunczaninovii (Drob.) Nevski var.macrathera Ohwi; Row 6: Roegneria varia Keng; Roegneria kamoji Ohwi; Row 7: Stipa grandis, Leymus chinensis (Trii) Tzvel.)

References

Bian, Z. Q., and X. G. Zhang. 2000. Pattern recognition (2nd Version)[M]. Beijing: Tsinghua University Press. (in Chinese).

Cai, C., M. Zhang, and J. P. Zhu. 2010. Weed seeds classification based on compressive sensing theory[J], Science China: Inforamtion Science, 40(Supplement): 160-172. (in Chinese).

Cui, D. F. 2010. Plant taxonomy(3rd Version)[M]. Beijing: Chinese Agriculture Press. (in Chinese).

Cui, J. 2014. 2D and 3D palmprint fusion and recognition using PCA plus TPTSR method[J]. Neural Comput& Applic, 24(3-4):497-502.

Duda, R. O., P. E. Hart, and D. G. Stork. 2001. Pattern Classification(2nd Edition)[M]. U.S: John Wiley & Sons.

Gonzalez, R. C., and R. E. Woods. 2008. Digital Image Processing[M]. Prentice Hall.

Han, J. G., and P. S. Mao. 2011. Forage seed science (2nd Version)[M]. Beijing: China Agricultural University Press. (in Chinese).

Ho, H. T., and R. Chellappa. 2013. Pose-Invariant face recognition using markov random fields. IEEE Transactions on Image Processing, 22(4):1573–1584.

Hong, P., T. Hai, L. Lan, V. Hoang, V. Hai, and T. Nguyen. 2015. Comparative study on vision based rice seed varieties identification[C], 7th International Conference on Knowledge and Systems Engineering. 377–382.

Granitto, P. M. , Navone, H. D. , Verdes, P. F. , & Ceccatto, H. A. . 2002. Weed seeds identification by machine vision[J], Computers and Electronics in Agriculture, 33:91-103.

Granitto, P. M., Verdes, P. F., Ceccatto, H. A., 2005. Large-scale investigationof weed seed identification by machine vision. Computers and Electronics in Agriculture. 47,15-24.

Pourreza, A., H. Pourreza, M. Abbaspour-Fard, and H. Sadrnia. 2012. Identification of nine Iranian wheat seed varieties by textual analysis with image processing[J]. Comput. Electron. Agr. 83:102-108.

Shi, C., and G. Ji. 2009. Study of recognition method of leguminous weed seeds image[C]. in: Proceedings of International Workshop on Intelligent Systems and Applications, 1–4.

Wang, N., Q. Li, A. El-Latif, J. Peng, and X. Niu. 2014. An enhanced thermal face recognition method based on multiscale complex fusion of Gabor coefficients[J]. Multimed. Tools Appl.72:2339-2358.

Zhang, D. Automated Biometrics-Technologies and Systems[M]. Kluwer Academic Publishers. 2000.

Zhang, D. Palmprint authentication[M]. Kluwer Academic publishers, 2004.

Chapter 3
Identification of Gramineous Grass Seeds Using Gabor and Locality Preserving Projections

Therefore, considering the importance of gramineous grass in grassland forage and urban landscaping, we propose a classification system of gramineous grass based on seed images using Gabor filter and locality preserving projections (Pan et al. 2016). The classification system comprises 4 modules: image acquisition, image preprocessing, feature extraction and feature matching (Fig. 3.1). At first, the seed images are captured by a common digital camera with a macro lens of 100 mm under daylight illumination. Then morphological operations are used for noise elimination and the Region of interest (ROI) segmentation. In the third module, Gabor wavelets and LPP are applied for robust feature extraction. The former can provide robust features against varying image brightness and contrast (Zhang et al. 2003), and the latter can preserve the image manifold structure (He et al. 2005) for efficient dimensionality reduction. The integration yields robust features for seed image presentation. Finally, the nearest neighbor (NN) classifier with Euclidean distance, and linear discriminant analysis (LDA) classifier are used for classification. The experimental results demonstrate the effectiveness of the proposed algorithm. The novelty lies in applying Gabor filters and LPP to extract the textural manifold features of the seed images, not limited to the appearance and geometric features.

Therefore, in the proposed system in this study, Gabor filters and locality preserving projections (LPP) are attempted to extract textural manifold features of gramineous seeds for forage identification.

In this chapter, we introduce a classification algorithm of gramineous grass based on seed images using Gabor filter and locality preserving projections (LPP). It is a feature extraction module in the recognition system, executed after image acquisition and preprocessing that were introduced in Sect. 3.2 (Fig. 3.1). The former Gabor wavelets can provide robust features against varying image brightness and contrast (Zhang et al. 2003), and the latter LPP can preserve the image manifold structure (He et al. 2005) for efficient dimensionality reduction. The integration yields robust features for seed image presentation. Finally, the nearest neighbor (NN) classifier with Euclidean distance, and linear discriminant analysis (LDA) classifier are used for classification. The experimental results demonstrate the effectiveness of the proposed

X. Pan et al., *Computer Vision based Identification and Mosaic of Gramineous Grass Seeds*, https://doi.org/10.1007/978-981-16-3501-4_3

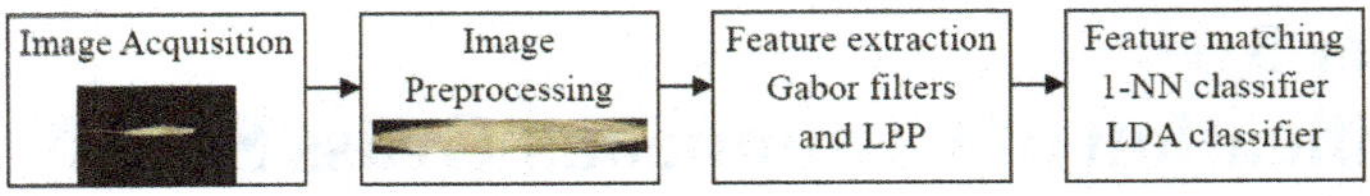

Fig. 3.1 Block diagram of the identification system for gramineous grass seeds

algorithm. The novelty lies in applying Gabor filters and LPP to extract the textural manifold features of the seed images, not limited to the appearance and geometric features.

In the remainder of this chapter, we will predominantly focuses on the feature extraction with Gabor and LPP, and feature matching with NN and LDA in Sect. 3.1. The experimental results and discussions are listed in Sect. 3.2. Section 3.3 highlights the conclusions.

3.1 Feature Extraction

In order to extract the robust features of the gramineous seeds images effectively, Gabor filters and LPP are integrated for manifold texture extraction against variations, especially illumination variations of the seed images. The application in face recognition shows the robust characteristics in texture analysis (Hu 2012).

3.1.1 Gabor Filters

The imaging condition has a significant impact on the classification efficiency (Pourreza et al. 2012). In consideration of the sophisticated surroundings of image acquisition procedure of the proposed system, such as non-controlled illumination under daylight, auto flash used for light compensation, automatic focus, and hand-held camera without mounting facilities, Gabor filters were used to alleviate the effects caused by imaging condition before further processing. Numerous studies have demonstrated that Gabor wavelets can provide robust features against varying image brightness and contrast in addition to accurate time frequency location. The successful application of Gabor filters in texture analysis involves a large quantity of areas, such as iris recognition, face recognition, etc. The utility of Gabor filters in the separation of weed and crops from natural images proves the good performance on weed texture (Ishak et al. 2009).

2D Gabor filters exhibits the following general form (Daugman 1980):

$$G(x, y, \theta, u, \sigma) = \frac{1}{2\pi\sigma^2}\exp\left\{-\frac{x^2 + y^2}{2\sigma^2}\right\}\exp\{2\pi i(ux\cos\theta + uy\sin\theta)\} \quad (3.1)$$

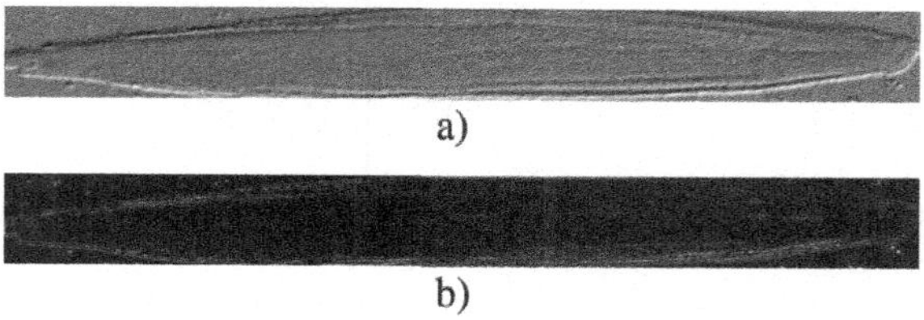

Fig. 3.2 Gabor filtered image, **a** imaginary part, **b** amplitude

where $i = \sqrt{-1}$, u is the frequency of the sinusoidal wave, θ controls the orientation of the function, and σ is the standard deviation of the Gaussian envelope. For an input seed image $I(x, y)$, the convolution of Gabor filter and the seed image yields Gabor filtered images $O(v, k)$. The transformed results include real and imaginary parts, which can be used to calculate the amplitude. Let OR and OI represent the real and imaginary parts, respectively, the magnitude and phase can be computed by the following equation:

$$O_{\text{amp}} = \sqrt{O_R^2 + O_I^2} \tag{3.2}$$

The imaginary part and the amplitude of the Gabor features are displayed in Fig. 3.2a, b, respectively. For N training images, all of the Gabor-filtered vectors form the Gabor feature space GS = $\{O_1, O_2, \ldots,$ ON$\}$.

3.1.2 Locality Preserving Projections

Locality preserving projections, an appearance-based linear subspace method, preserve the intrinsic geometry of the data and local manifold structure modeled by a nearest neighbor graph of the images (He et al. 2005; Cai et al. 2006). In the proposed system, LPP is applied onto the Gabor feature space GS of the seed images. The integration can provide robust features for varying brightness and image contrast while preserving the manifold structure of the images for efficient dimensionality reduction. The following specific steps are listed:

(1) Principal component analysis (PCA) projection. The Gabor feature space is first projected onto the PCA subspace, and N principle components are preserved because the number of training samples is relatively small compared to the dimension of training vectors.
(2) Construction of the nearest neighbor graph and the similarity matrix *S*. When each node corresponds to a Gabor filtered vector, the nearest neighbor graph of *N* nodes can be constructed. For a node *i*, it can be connected with node *j* if they are close, where their distance is among the *k* nearest one of node *i* and all other nodes in the graph. The similarity matrix *S* can be defined as an $N \times N$ matrix,

$$S = \begin{bmatrix} S(11) & S(12) & \cdots & S(1N) \\ S(21) & S(22) & \cdots & S(2N) \\ \vdots & \vdots & \cdots & \vdots \\ S(N1) & S(N2) & \cdots & S(NN) \end{bmatrix} \quad (3.3)$$

where $s(ij) = 1$, if node j is among k nearest neighbors of i, otherwise $s(ij) = 0$, $i, j = 1, 2, \ldots, n$.

(3) Eigenmap: Computation of the optimum LPP transform matrix by solving the following generalized eigenvalues problem:

$$ZLZ^T a = \lambda ZDZ^T a \quad (3.4)$$

where D is a diagonal matrix whose elements are defined as $D(ii) = \sum_{j=1}^{N} s(ji)$. $L = D - S$ is the Laplacian matrix. The minimum eigenvectors $a_1, \ldots, a_s$ of Eq. (3.5) corresponding to the s minimum eigenvalues form the optimal projection matrix A_{opt}, i.e., $A_{\text{opt}} = [a_1, \ldots, a_s]$.

Once the optimum projection matrices $A_{\text{opt}} = [a_1, \ldots, a_s]$ is obtained, we can project the Gabor feature vector $X(i)$ onto A to yield a s-dimensional feature matrix $Y(i)$

$$Y(i) = A^T X(i) \quad (3.5)$$

Similarly, for a test image $X(\text{test})$, the feature matrix $Y(\text{test})$ is

$$Y(\text{test}) = A^T X(\text{test}) \quad (3.6)$$

3.2 Feature Matching

Feature matching is another key module that affects the identification results. In this study, we choose the nearest neighbor classifier (1-NN) and Linear Discriminant Analysis (LDA) classifier for efficient and discriminant matching, respectively.

The nearest neighbor classifier is a simple and commonly used tool for identification. The test image U is classified as belonging to class P if their distance (such as Euclidean distance, Cityblock distance, Cosine distance and Correlation distance, etc.) is the minimum among all the classes in the database. The above statement can be expressed as follows, if

$$d(U, P) = \min_T d(U, T), T = 1, 2 \ldots, c, \quad (3.7)$$

then decide. $U \in P$

Linear Discriminant Analysis (LDA), also known as the Fisher discriminant, is a supervised classification method that uses both statistical and geometrical measurements to determine which group an unknown data point belongs to (http://cn.mathworks.com/help/stats/discriminant-analysis.html). It is expected to gain more information about within class and between classes for discriminant results in the study.

3.3 Results and Discussion

To testify the effectiveness of the algorithms, two image databases of gramineous grass seed were constructed, in which the number of seeds species, and acquisition approach were different. The seeds, which are listed in Table 2.1 by name and origin, were provided by the Grassland Research Institute of the Chinese Academy of Agricultural Sciences. Hence, we discussed the experimental results executed on the two database in Sects. 4.1 and 4.2, respectively. All of the experiments were executed on an Intel Core i5-2467M CPU @ 1.60 GHz and 6 GB RAM; the codes were written in MATLAB 2011b.

3.3.1 Experiments on Seed Image DatabaseI

The two groups of experiments according to two practical cases were executed in Sects. 3.3.1 and 3.3.2. In the first case, a seed had images in both training set and test set, similar to face identification of a person who has training samples in the database. The purpose corresponds to the identification of a seed already existed in the database. However, most seeds have no training samples in the database, and the identification would tell the seed category in practical circumstances. Hence, in the second case, we kept images of the same seed in the database separate in either training set or test set to simulate the real circumstance. The identification accuracy (or correct identification rate) is an indicator of identification performance, defined as the ratio of the numbers that are correctly identified and the total test samples.

3.3.1.1 Identification of Gramineous Grass Seeds with Images in Both Training Set and Test Set

In the first case of identification experiments, every seed had samples both in training and test set. Appearance-based algorithm (Granitto et al. 2002), LPP, and Gabor-based LPP were executed on the seed base to test the effectiveness of the proposed algorithm. Euclidean distance and nearest neighbor classifier are used to measure and classify similarity. For simplicity, the All of the ROIs in LPP related algorithm were resized to a uniform resolution of 25×250 and then vectored into a column. The

ROIs in appearance-based algorithm were not normalized to the same size for the intrinsic geometry features required. To reduce the cost of computation and memory, only one Gabor filter with the following parameters was applied, referring to the previous experience of texture analysis with Gabor filters (Pan and Ruan 2008). Thus, the dimension of the Gabor space formed were 6250 × N, where N represents number of training samples. For LPP, N principle components were preserved in PCA step because the number of training samples was relatively small as compared to the dimension of training vector dimension. To eliminate the effect of choosing samples randomly, K-fold cross-validation was employed in all the following experiments.

A. **Identification performance with 80% training samples and 20% testing samples.**

In the first sub-group of experiments, we compared correct classification rates of the three algorithms using 4 images per seed as training samples and the remaining 1 image as the testing sample, i.e., each class comprised 40 samples for training and 10 samples for testing. The training set and the testing set contained 560 and 140 images, respectively. Table 3.1 listed the correct classification rates using different testing samples from No. 1 to No. 5 per seed and their average, comparing LPP based on original images and Gabor features (imaginary part, real part and magnitude). The corresponding feature dimensions are listed in parentheses below. In appearance-based algorithm (Granitto et al. 2002), the feature dimension kept a constant of 12, composed of selected 6 morphological, 4 color and 2 textural seed characteristics. In LPP and Gabor based LPP, the identification rate varied with the feature dimension, and the feature dimensions listed corresponded to the top identification rates in the experiments.

It can be observed from Table 3.1, the identification rates varied with the combinations of different training and testing samples, which proved that the imaging quality of different seed images had an effect on the identification performance of all the algorithms. Hence, it was an effective way to calculate the average identification rate to decrease the relative error caused by choosing samples.

Table 3.1 Comparison of the correct recognition rate (%) and feature dimension using K-fold cross validation (40 training samples per class, 10 testing samples per class)

Testing sample no	No. 1	No. 2	No. 3	No. 4	No. 5	Average
LPP	92.86	97.86	97.14	97.86	97.14	96.57
	(24)	(113)	(48)	(53)	(44)	
Gabor (imaginary part) + LPP	95.71	99.29	98.57	100	99.29	98.57
	(34)	(22)	(39)	(32)	(30)	
Gabor (real part) + LPP	95.71	99.29	97.86	100	98.57	98.29
	(37)	(49)	(35)	(24)	(52)	
Gabor (magnitude) + LPP	94.29	100	97.14	100	99.29	98.14
	(65)	(68)	(29)	(47)	(70)	

The average LPP correct identification rate reached 96.57%. The top classification rate of 97.86% was obtained when No. 2 or No. 4 images were used as the testing sample per seed. Gabor features improved the identification rates of LPP no matter which part used, real, imaginary, and the amplitude. This observation is mainly attributed to the robustness of the Gabor features in eliminating the effect of illumination variation. The highest correct identification rate of 100% occurred with No. 4 as the testing sample, with the feature dimension less than 47, which suggested high classification efficiency in the feature matching stage. By comparing the average correct identification rates, the Gabor-based LPP with the imaginary part exhibited the highest classification performance. The average correct classification rate reached 98.57, 2% higher than LPP's alone. Therefore, the Gabor features in the remainder of the experiments denoted the imaginary part of Gabor filtered images.

As the identification accuracies vary with the feature dimension of the LPP associated algorithms, Fig. 3.3 illustrates the identification accuracies under different dimensions using the No. 4 sample per seed for testing. The horizontal axis and vertical axis represent feature dimensions and the correct identification rates, respectively. The Gabor-based LPP is more stable and yields a better performance than that of LPP, which suggested that adopting the Gabor features improved the identification performance. This is consistent with the conclusion drawn from most studies that Gabor features are more robust against variations in image illumination and contrast. The correct identification rates of the two algorithms increased rapidly as

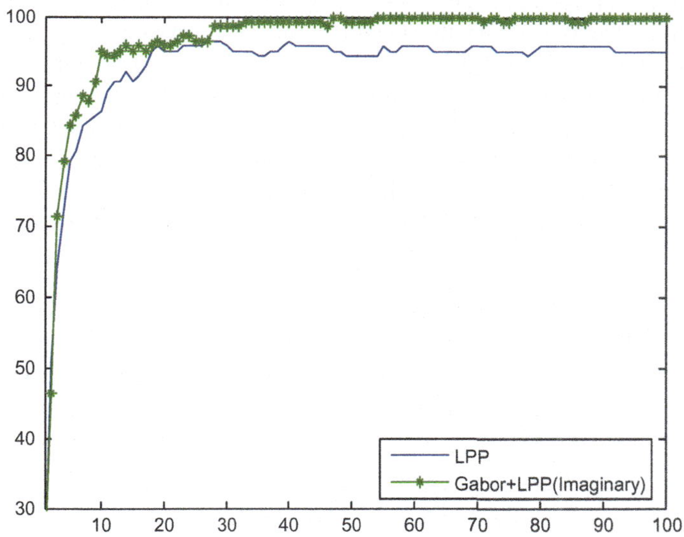

Fig. 3.3 Comparison of correct identification rates of LPP and Gabor-based LPP (imaginary part) under different dimensions

the dimensionality of the subspace increased, partly because they both used LPP for dimensionality reduction. However, the identification accuracy of Gabor-based LPP increased much more rapidly after dimension 3 (71.42% Gabor based LPP, 64.28% LPP) and achieved the best result with 100% (dimension 47), 5% higher than the identification rate of LPP based on original images. By comparing the graph, the proposed Gabor-based LPP outperformed LPP and exhibited a consistent superiority from dimension 3. Hence, Gabor-based LPP showed a good consistency in identification performance when using a relatively large quantity of training samples.

B. **Identification performance with 80% training samples and 20% testing samples.**

The second sub-group of experiments was conducted to test the identification accuracies of the above three algorithms using one training sample per seed. Thus one hundred and forty training samples and 560 testing samples were used in the experiments, i.e. the ratios of training samples and testing samples are 80% and 20%, respectively. All possible training sample and testing sample combinations were tested to eliminate the effect of random sample bias.

Table 3.2 listed the identification rates and their corresponding dimensions in parentheses. The feature dimensions listed correspond to the top identification rates in the LPP executed on original image and Gabor features. As can be seen from Table 3.2,

LPP based on original images also decreased from 96.57 (Table 3.1) to 75.54%, demonstrating the impact on the identification performance using less training samples and more testing samples on the manifold subspace algorithm. Gabor based LPP was less affected by the variations of training samples and testing samples, with an average identification of 87.32, 11.25% less than the identification rate 98.57% in Table 3.3. The top identification accuracy reached 89.11% with a dimension of 64. The experimental results suggested that the integration of Gabor features and LPP enhanced the performance of classification when the number of training samples

Table 3.2 Comparison of the correct recognition rate (%) and feature dimension using K-fold cross validation (10 training samples per class, 40 testing samples per class)

Training samples No.	Testing samples no	Gabor + LPP (imaginary)	LPP
No. 1	No. 2, 3, 4, 5	83.57 (70)	71.79 (88)
No. 2	No. 1, 3, 4, 5	88.21 (63)	76.79 (82)
No. 3	No. 1, 2, 4, 5	88.04 (67)	76.43 (59)
No. 4	No. 1, 2, 3, 5	89.11 (64)	76.43 (93)
No. 5	No. 1, 2, 3, 4	87.68 (65)	76.25 (100)
Average		87.32	75.54

Table 3.3 Comparison of the correct identification rates (%) of appearance-based algorithm

	Euclidean + 1-NN	Cityblock + 1-NN	Cosine + 1-NN	Correlati-on + 1-NN	Bayes	LDA
No. 1	24.2857	24.2857	18.5714	25.7143	20.0000	**52.8571**
No. 2	24.2857	21.4286	22.8571	21.4286	31.4286	37.1429
No. 3	31.4286	31.4286	30.0000	31.4286	21.4286	42.3571
No. 4	37.1429	34.2857	34.2857	35.7143	21.4286	42.8571
No. 5	17.1429	15.7143	17.1429	18.5714	28.5714	38.5714
No. 6	24.2857	21.4286	22.8571	35.7143	27.1429	48.5714
No. 7	31.4286	27.1429	28.5714	30.0000	27.1429	20.0000
No. 8	38.5714	41.4286	34.2857	32.8571	30.0000	50.0000
No. 9	28.5714	28.5714	27.1429	21.4286	21.4286	34.2857
No. 10	35.7143	30.0000	27.1429	32.8571	24.2857	44.2857
Ave	29.28572	27.57144	26.18571	28.57143	**25.28573**	**41.09284**

was small. The Gabor filtered LPP had a better performance than LPP based on gray images using different training samples which verified the robustness of Gabor features.

3.3.1.2 Identification of Gramineous Grass Seeds with Images in Either Training Set or Test Set

Different from biometrics of human being, the samples to be tested in plant classification have no training images in most cases, for it is impossible to collect all the samples of one species in the world into the database. The classification of unknown samples into a certain category depends on the knowledge of the existed samples. Hence, it becomes a much more difficult task due to the individual variations of the same species, and similarities in different species, especially in one family, which commonly exists in biological world. Hence, in the second case of identification experiments, every seed to be tested had no training samples in the training set, i.e. the seeds in the training set and test set are not same.

Since each species in the database was represented by 50 images captured from 10 seeds, 5 images per seed, we grouped 45 images of 9 seeds and the rest 5 images of 1 seed into training set and test set, respectively. In order to eliminate the effect of choosing samples, 10 seeds of each class was used as testing samples and the rest were used for training in turn. Thus the total number of training set and test set are 630 and 70 for all 14 species, respectively. Figure 3.4 showed the first 10 PCA eigenvectors (in the left column) and Laplacian eigenvectors (in the right column) of the gramineous seeds when using 45 images of the first 9 seeds as training set. The images may be called Gabor based Eigenseeds and Laplacianseeds, respectively. As an intermediate step of LPP, PCA effectively see only the Euclidean structure. Comparatively, LPP

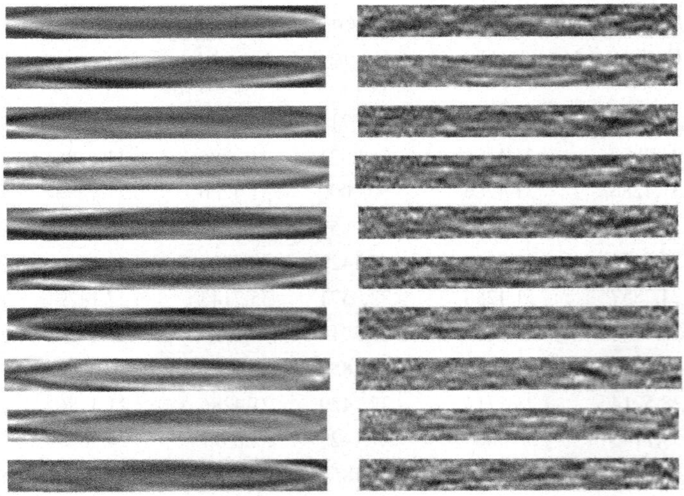

Fig. 3.4 The first Gabor-based 10 Eigenseeds (left column), and Laplacianseeds (right column) calculated from the gramineous seed database

preserves local manifold structure modeled by a nearest-neighbor graph (He et al. 2005). A seed image can be mapped into the locality preserving subspace by using the Laplacianseeds.

Several feature extraction algorithms including appearance-based (Granitto et al. 2002), local binary Pattern (LBP), and Gabor-based LPP algorithms were executed for comparison. LBP approach is a texture descriptor method gained widely acceptance due to its computational simplicity and powerful performance for describing textures (Nava et al. 2011). Experimental results listed in Tables 3.3, 3.4 and

Table 3.4 Comparison of the correct identification rates (%) of LBP algorithm

	Euclidean + 1-NN	Cityblock + 1-NN	Cosine + 1-NN	Correlati-on + 1-NN	Bayes	LDA
No. 1	20.0000	34.2857	27.1429	25.7143	24.2857	40.0000
No. 2	21.4286	30.0000	27.1429	27.1429	14.2857	37.1429
No. 3	22.8571	27.1429	24.2857	24.2857	17.1429	38.5714
No. 4	25.7143	30.0000	22.8571	22.8571	28.5714	40.0000
No. 5	31.4286	30.0000	27.1429	27.1429	30.0000	50.0000
No. 6	27.1429	32.8571	30.0000	30.0000	17.1429	44.2857
No. 7	21.4286	21.4286	24.2857	28.5714	18.5714	40.0000
No. 8	30.0000	32.8571	32.8571	32.8571	28.5714	**51.4286**
No. 9	12.8571	11.4286	14.2857	14.2857	21.4286	28.5714
No. 10	35.7143	38.5714	34.2857	35.7143	20.0000	40.0000
Ave	24.85715	28.85714	26.42857	26.85714	22.0000	**41.0000**

Table 3.5 Comparison of the correct identification rates (%) of Gabor based LPP algorithm

	Euclidean + 1-NN	Cityblock + 1-NN	Cosine + 1-NN	Correlati-on + 1-NN	NBC	DAC
No. 1	41.4286 (35)	35.7143 (45)	47.1429 (35)	47.1429 (34)	51.4286 (24)	50.0000 (28)
No. 2	45.7143 (56)	48.5714 (5)	50.0000 (42)	52.8571 (28)	55.7143 (50)	55.7143 (57)
No. 3	50.0000 (20)	47.1429 (20)	50.0000 (28)	47.1429 (92)	65.7143 (7)	54.2857 (21)
No. 4	54.2857 (29)	54.2857 (27)	57.1429 (62)	55.7143 (52)	55.7143 (11)	67.1429 (28)
No. 5	41.4286 (74)	37.1429 (12)	51.4286 (23)	57.1429 (26)	45.7143 (81)	51.4286 (68)
No. 6	51.4286 (12)	48.5714 (85)	65.7143 (25)	62.8571 (24)	71.4286 (44)	**80.0000** (29)
No. 7	41.4286 (15)	47.1429 (20)	54.2857 (11)	42.8571 (12)	57.1429 (98)	60.0000 (106)
No. 8	55.7143 (72)	54.2857 (49)	61.4286 (62)	68.5714 (62)	65.7143 (89)	71.4286 (101)
No. 9	47.1429 (23)	50.0000 (28)	58.5714 (29)	60.0000 (32)	61.4286 (12)	62.8571 (41)
No. 10	60.0000 (31)	55.7143 (30)	71.4286 (21)	68.5714 (37)	67.1429 (14)	82.8571 (20)
Ave	48.85716	47.85715	56.7143	56.28571	59.71431	**63.57143**

3.5 corresponded to the correct identification rates of appearance-based algorithm (Granitto et al. 2002), LBP and Gabor-based LPP algorithms, respectively. Considering the impact on the identification performance caused by classification method, the Nearest Neighborhood Classifier (NNC) integrated with Euclidean distance, Cityblock distance, Cosine distance, and Correlation distance, Naïve Bayes Classifier (NBC), and Linear Discriminant Analysis (LDA) were compared for the optimum identification performance as well. The feature dimensions were not listed in Tables 3.3 and 3.4 because they kept constants of 12 and 256, respectively. The 12 feature components of appearance-based algorithm were composed of 6 selected morphological, 4 color and 2 textural characteristics (Granitto et al. 2002). The 256 feature components in LBP came from LBP histogram, treated as a fixed-dimensional vector for distance evaluation (Nava et al. 2011). The feature dimensions of Gabor based LPP were listed in the parentheses corresponding to the top correct identification rates in Table 3.5.

As can be seen, the overall identification accuracies were not as high as that of experiments in Sect. 3.3.1.1. The average identification rates of appearance-based algorithm (Granitto et al. 2002) dropped from 72.572% (Table 3.1) to 41.09284% (Table 3.3). Similarly, the average identification rates of Gabor based LPP dropped from 98.57% (Table 3.1) to 63.57143% (Table 3.5). The obvious decreases occurred when the test seeds have no training samples in the database. It was the individual variations within the same species and the similarity of different species in gramineous grass family that made the identification task more complicated. The impact of individual varieties within the same species on identification performance can also be witnessed by the obviously varied identification accuracies when using 10 seeds in the same species as test images in turn (Tables 3.3, 3.4 and 3.5).

Appearance-based algorithm had a top average identification accuracy 41.09284% when using LDA classifier, as shown in Table 3.3. Bayes Classifier showed the worse identification performance with an average accuracy of 25.28573%, suggesting that the combination of selected features and Bayes classifier (Granitto et al. 2002) was only effective for certain weed seeds, rather than gramineous grass seeds. The top identification rate LBP reached 41% using LDA classifier, as shown in Table 3.3, 12.14286% higher than 28.85714% tested by Cityblock distance and NNC. Despite the descriptive ability of texture, LBP failed to reveal a higher identification accuracy mainly because it cannot well represent the intrinsic manifold structure of the seed surface. The top average identification accuracy of Gabor based LPP was 63.57143% using LDA classifier (Table 3.5), almost 22.48% and 22.57% higher than that of appearance-based algorithm and LBP, respectively. When choosing the images of Seed No. 10 as the test samples, the correct identification rate reached 82.8571% with a feature dimension of 20. The experimental results demonstrated the robustness of textural manifold features extracted from images of gramineous grass seeds. Moreover, LDA Classifier was beneficial to the improvement of the discriminant ability of textural manifold features in classification.

3.3.2 Experiments on Seed Image DatabaseII

To enhance the practicality of the system, the second gramineous grass seed image DatabaseII was constructed by taking photos of 6 seeds at a time outside the lab (Fig. 3.4). The database contained 300 seeds from 6 species of gramineous grass (Elymus nutans Griseb, Elymus cylindricus (Frmch) Hinda, Agropyron cristatum var. pectiniforme (Roem.et Schult) H Yang, Bromus inermis Leyss., Elymas sibiricus Linn, Roegneria kamoji Ohwi). Each species has 50 seeds, and each seed has one mere image. The ROIs were cropped individually for further feature extraction. Figure 3.5 listed some ROI examples of the listed species (Fig. 3.6).

Feature extraction algorithms including appearance-based (Granitto et al. 2002), local binary Pattern (LBP), and Gabor-based LPP algorithms together with different classifiers, such as Nearest Neighborhood Classifier (NNC) integrated by Euclidean distance, Cityblock distance, Cosine distance, and Correlation distance, Naïve Bayes Classifier (NBC), and Linear Discriminant Analysis (LDA) were executed on SeedBase II, similar as experiments in Sect. 3.3.1.2. The test set was grouped by randomly choosing 5 samples of each species, and the rest samples were used for training. Since each seed had one image, every seed to be tested had no training samples in the training set, similar as the experiments in Sect. 3.3.1.2. Table 3.6 listed the correct identification rates of different algorithms and classifiers.

Fig. 3.5 Original seed image (Elymas sibiricus Linn) (databaseII)

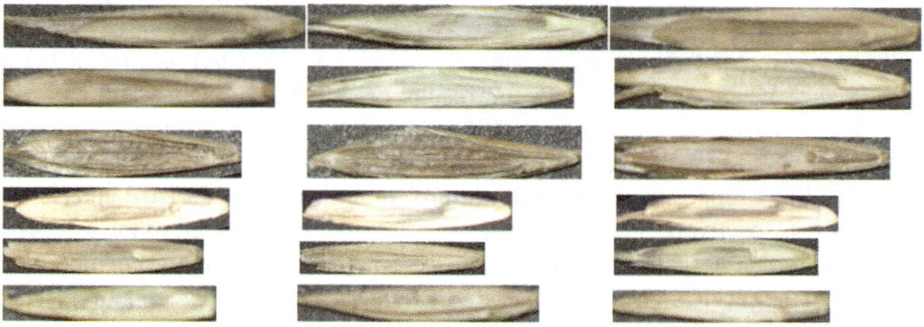

Fig. 3.6 Seed image databaseII of Gramineous Grass (from top to bottom. Row 1: Roegneria kamoji Ohwi; Row 2: Elymas sibiricus Linn; Row 3: Bromus inermis Leyss.; Row 4: Elymus cylindricus (Frmch)Hinda; Row 5: Elymus nutans Griseb; Row 6: Agropyron cristatum var.pectiniforme (Roem.et Schult)H Yang

Table 3.6 Comparisons of the correct identification rates (%) of different algorithms (SeedbaseII)

Algorithms	Euclide-an + NNC	Cityblo-ck + NNC	Cosine + NNC	Correlati-on + NNC	NBC	LDA
Appearance (Granitto et al. 2002)	40.00 (12)	33.33 (12)	26.67 (12)	30.00 (12)	**53.33 (12)**	**53.33 (12)**
LBP	46.67 (256)	**56.67 (256)**	46.67 (256)	46.67 (256)	36.67 (256)	16.67 (256)
Gabor + LPP	66.67 (30)	63.33 (31)	70.00 (66)	73.33 (29)	80.00 (23)	**86.67 (18)**

The feature dimensions of appearance-based algorithm and LBP kept constants of 12 and 256, respectively. The 12 feature components of appearance-based algorithm includes 6 selected morphological, 4 color and 2 textural characteristics (Granitto et al. 2002). The 256 feature components in LBP came from LBP histogram, treated as a fixed-dimensional vector for distance evaluation (Nava et al. 2011).The feature dimensions of Gabor based LPP were listed in the parentheses corresponding to the top correct identification rates. As can be seen, appearance-based algorithm attained a top identification accuracy 53.33% when using LDA classifier and Bayes Classifier, suggesting the two classifiers has a better discriminate ability when the number of samples are relatively sufficient. LBP presented an overall better performance than appearance-based algorithm and has the top identification accuracy 56.67% using the Nearest Neighbor Classifier integrated with Cityblock. The top average identification accuracy of Gabor based LPP was 86.67% using LDA classifier, almost 30.00% and 33.34% higher than that of appearance-based algorithm and LBP, respectively. The corresponding feature dimension 18 indicated an efficient matching speed as compared with 256 of LBP. The experimental results demonstrated the robustness of textural manifold features extracted by Gabor based LPP from images of gramineous grass seeds.

The experimental results on SeedbaseII (Table 3.6) revealed an overall better performance when compared with the results on SeedbaseI (Tables 3.3, 3.4 and 3.5). The top identification rates 53.33%, 56.67% and 86.67% (Table 3.6) were higher than 52.8571% (Table 3.3),51.4286% (Table 3.4), and 82.8571% (Table 3.5), corresponding to appearance-based algorithm (Granitto et al. 2002), LBP, and Gabor-based LPP, respectively. The improvements were mainly attributed to the relatively sufficient number of seed samples 50 versus 10, which can better represent biological diversity of the species. The Gabor-based LPP outperformed all the competitors listed in experimental results by a large margin, suggesting the manifold textural features were fit for the identification of gramineous grass seeds.

3.4 Summary

In this study, we presented an identification system of gramineous grass based on seed images. The classification system comprises four modules: image acquisition, preprocessing, feature extraction and matching. In the image acquisition module, we constructed two gramineous grass seed image databases, which are expected to be a public platform for future research into grass classification. In the image-preprocessing module, an opening mathematical morphology operation was used for ROI segmentation. The integration of Gabor filters and LPP was used in the feature extraction module to guarantee the identification accuracy, for Gabor features are robust under variable illumination and surroundings, and LPP was an effective manifold learning approach to reduce the Gabor feature dimensions while preserving the intrinsic structure. The novelty of the system of the algorithm was to extract robust manifold structures rather than in the appearance and geometrical features. The nearest neighbor classifier integrated with Euclidean distance, and LDA classifiers were used in the feature matching for efficient classification. Gabor-based LPP outperformed appearance-based algorithm and LBP when using both seed image databases, indicating that grass classification can be realized based on seed images by extracting textural manifold features.

Future work will explore other approaches to the identification of forage, such as assessing the leaves or flowers. In addition, more images from different grassland species should be collected to validate the robustness and enhance the practicality of the grass classification system.

References

Cai, D., X. He, J. Han, and H. Zhang. 2006. Orthogonal Laplacianfaces for face recognition[J]. Trans. IEEE Image Processing. 15:3608–3614.

Cui, J. 2014. 2D and 3D palmprint fusion and recognition using PCA plus TPTSR method[J]. Neural Comput& Applic. 24:497–502.

Daugman, J.G. 1980. Two-dimensional spectral analysis of cortical receptive field profiles[J]. Vision Res. 20:847–856.

Granitto, P. M., H. D. Navone, P. F. Verdes, and H. A. Ceccatto. 2002. Weed seeds identification by machine vision[J]. Computers and Electronics in Agriculture. 33:91–103.

He, X., S. Yan, Y. Hu, P. Niyogi, and H. Zhang. 2005. Face recognition using Laplacianfaces. [J] Trans. IEEE Pattern Anal. Mach. Intel. 27: 328–340.

Hu, P. 2012. Application research of Gabor filter and LPP algorithms in face recognition[J]. Lecture notes in electrical engineering. 144:483–489.

Ishak, A. J., A. Hussain, and M. M. Mustafa. 2009. Weed image classification using Gabor wavelet and gradient field distribution[J]. Computers and Electronics in Agriculture. 66:53–61.

Nava, R., G. Cristobal, and B. Escalante-Ramirez. 2011. Invariant texture analysis through local binary patterns[J], Pattern Recognition letters, 1–30.

Pan, X., and Q. Ruan. 2008. Palmprint recognition with improved two-dimensional locality preserving projections. Image and Vision Computing. 26(9):1261–1268

Pan, X., Y. Cen, Y. B. Ma, and W. H. Yan. 2016. Identification of gramineous grass seeds using Gabor and locality preserving projections [J]. Multimedia Tools and Applications. 75(23):16551–16576.

Pourreza, A., H. Pourreza, M. Abbaspour-Fard, and H. Sadrnia. 2012. Identification of nine Iranian wheat seed varieties by textual analysis with image processing [J]. Comput. Electron. Agr. 83:102–108.

Zhang, D., W. K. Kong, J. You, and M. Wong. 2003.Online palmprint identification. Trans[J]. IEEE Pattern Anal. Mach. Intel. 25, 1041–1050.

Chapter 4
Identification of Gramineous Grass Seeds Using Difference of Local Fractal Dimensions

When we focus on seed identification and quality assessment of forage, it is interesting to note that similar seeds, especially in some closely species of the same family (genus), are hard to be identified correctly, even for the experts. For example, gramineous grass, an important category in grassland forage and urban landscaping has some similar varieties, such as Agropyron cristatum var. pectiniforme (Roem.et Schult) H Yang (Fig. 4.1), Agropyron cristatum (Linn.) Gaertn (Fig. 4.2), Agropyron desertorum (Fisch.) Schult. (Fig. 4.3) and Agropyron mongolicum Keng (Fig. 4.4), etc., categorized to Agropyron Gaertn Genus. It can be observed that the varieties of same family are too similar to distinguish from the appearance. Texture analysis can be used for reference in similar seed identification. Pourreza et al. attempted to extract textual features for similar varieties in wheat seed identification. They investigated 9 Iranian wheat seed varieties conducted on bulk sample images and extracted 131 textural features for identification including GLCM (gray level co-occurrence matrix), GLRM (gray level run-length matrix), LBP(local binary patterns), etc.. LDA (linear discriminate analysis) classifier was employed for classification using top selected features, yielding an average classification accuracy of 98.15% with top 50 selected features.

In texture analysis, fractal geometry is a powerful modeling tool, achieving interesting results in the description and discrimination of textures. The surface texture of seed have local self-similarity and fractal dimension can measure the roughness and self-similarity of the shape and texture. Fractal dimension is a digital measurement for the roughness of object's surface. Due to the difficulties in describing and distinguishing a whole image with a mere fractal dimension, difference of local fractal dimension is proposed for feature extraction. In order to highlight the differences and relations between the blocks, each block is divided into sub-blocks of the same size twice, and the fractal dimensions of all sub-blocks and their variances are calculated to be arranged in a column to form the feature vectors of the image. To be concrete, the algorithm can be divided into three steps. First, the ROI images cropped from the original seed image are divided into blocks of same size for local partitions, and the fractal dimensions of all the blocks are calculated. Then, based on the average fractal

X. Pan et al., *Computer Vision based Identification and Mosaic of Gramineous Grass Seeds*, https://doi.org/10.1007/978-981-16-3501-4_4

Fig. 4.1 Agropyron cristatum var.pectiniforme (Roem.et Schult) H Yang

Fig. 4.2 Agropyron cristatum (Linn.) Gaertn

Fig. 4.3 Agropyron desertorum (Fisch.) Schult

Fig. 4.4 Agropyron mongolicum Keng

dimension of all the blocks, the difference of local fractal dimension (DLFD) can be gained by subtracting the individual fractal dimension and the average, expanding the contrast of the self-similarity of the seeds. The above DLFDs of all blocks form into the feature vectors for seed image representation. Finally, we calculate the Euclidean Distances as the input of nearest neighbor classifier for classification. The main innovation of the algorithm lies in that it adopts the idea of partial shape dimension variance to enlarge the difference of texture roughness inside the image and improve the classification effect (Pan et al. 2016).

The rest of this paper is organized according to the sequence of identification as follows. Section 4.1 introduces the fractal dimension. Section 4.2 mainly focuses on the feature extraction stage, difference of local fractal dimension algorithms, and feature matching is also involved. Experimental results are listed in Sect. 4.3. Section 4.4 highlights the conclusion.

4.1 Fractal and Fractal Dimension

Fractal geometry of nature was first proposed by Mandelbrot in 1975 to describe the objects exhibiting the self-similarity at all scales, belonging to an important nonlinear discipline. Fractal means "irregular or fractured, derived from the Latin "Fractus" (Mandelbrot 1982). So far, there is no strict definition of fractals, and

Falconner's description (Falconer 1990) of fractals as complex sets with or partly with the following typical properties is generally accepted:

(1) Fine structure. The details at any small scale exist, or the details at any scale are revealed.
(2) Highly irregular. The entirety cannot be described by Euclidean geometry.
(3) Usually some kind of self-similar structure. They may be approximate or statistical.
(4) Generally, "fractal dimension" is defined in a certain way larger than its topological dimension.
(5) In most cases, it can be defined in a very simple way or generated by iteration.

Self-similarity and fractal dimension are two important characteristics of fractal.

The concept of fractal dimension (FD) is widely used for texture analysis and classification and results in good performance (Shanmugavadivu and Sivakumar, 2012; Goncalces and Bruno 2013). FD is a simple indicator of measuring the roughness and self-similarity of shape and texture. Mandelbrot defined fractal dimension D of A by the following equation.

$$D = \frac{\log(N_r)}{\log(^1/_r)} \tag{4.1}$$

For a bounded set A in Euclidean n-space, if A is the union of N_r non-overlapping distinct copies of itself each of which is similar to A scaled down by a ratio *r*, the set is self-similar. However, it is very difficult to compute *D* directly. Sarkar and Chaudhuri (1994) developed a simple, accurate and efficient algorithm named differential box counting (DBC) for the estimation of FD (Fig. 4.1). In their method, an image of size M × M is scaled down to a size s × s where s is an integer $^M/_2 \geq s > 1$, then ratio $r = {}^s/_M$. Considering the image as a 3-D space with coordinates (*x, y*) and (*z*) denoting 2-D position and gray level respectively, the image is partitioned into grids by the s × s × s boxes. If the minimum and the maximum gray level of the (*i, j*)th grid fall in the box number *k* and l respectively, the contribution of N_r in the (*i, j*)th grid is

$$n_r(i, j) = 1 - k + 1. \tag{4.2}$$

And the total contributions N_r of all the grids is (Fig. 4.5)

$$N_r = \sum_{i,j} n_r(i, j). \tag{4.3}$$

then the FD at a scale *r* can be computed by Eq. (4.1).

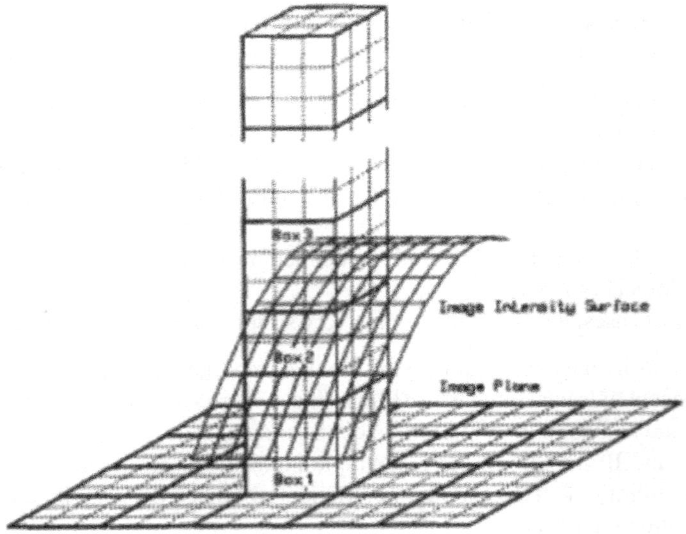

Fig. 4.5 Computation of N_r of differential boxing count (Sarkar and Chaudhuri 1994)

4.2 Difference of Local Fractal Dimension

Although fractal dimension excels at describing the richness texture of a complex image, and scale-dependent (Ishak et al. 2009), it is still hard to precisely represent a whole image with the only fractal dimension (Zou et al. 2007). Local matching partitioning the region of interest (ROI) into smaller sub-images can bind the effects of image variations and better preserve local information. Therefore, in this work, ROI seed image is divided into local partitions whose fractal dimensions are used for local information presentation. Considering that the fractal dimensions vary in a certain range, the differences of all blocks and their average form feature vectors to magnify the local variations.

Suppose an image *I* (*x*, *y*) of size M × N, the partition is to divide palmprint image into non-overlapping uniform subblocks of size 2 s × 2 s. Due to the contrast of minimum and the maximum gray level during calculation procedure of fractal dimension using BCD, the minimum block should be at least 22 × 22, that is s should be the integers between 2 and $\log_2 \min(M, N)$. Then the local fractal dimension of each block A_{pq} can be calculated as following

$$\mathrm{LFD}\left(A_{\mathrm{pq}}\right) = \sum_{i,j \in A_{\mathrm{pq}}} n_r(i, j) \tag{4.4}$$

where p and q are integers within the scope of $1 \leq p = {}^{M}/_{2^s}, 1 \leq q = {}^{N}/_{2^s}$. Considering that the values of the fractal dimension for all partitions are relatively close, the difference of the local fractal dimensions and their average are computed to magnify the contrast between local blocks. Meanwhile, the difference can alleviate the effect of illumination variation by subtracting the average fractal dimension. The average fractal dimension can be calculated as follows

$$\mu(\mathrm{FD}) = \sum_{p=1}\sum_{q=1} \mathrm{LFD}\left(A_{\mathrm{pq}}\right) \tag{4.5}$$

Thus the difference of the local fractal dimension for each partition is

$$\mathrm{DLFD}\left(A_{ij}\right) = \mathrm{FD}\left(A_{ij}\right) - \mu(\mathrm{FD}) \tag{4.6}$$

Then, we can form a feature vector composed of DLFD of all the subblocks for image representation

$$F_{\mathrm{DLFD}} = \left(\mathrm{DLFD}(A_{11}), \mathrm{DLFD}(A_{12}), \ldots, \mathrm{DLFD}\left(A_{M/2^s\, N/2^s}\right)\right) \tag{4.7}$$

Having obtained the feature vectors, the Euclidean distance and nearest neighbor classifier are used to measure and classify similarity for simplicity, respectively. The concrete equation of Euclidean distance is

$$d\left(F_{\mathrm{DLFD}^{(\mathrm{test})}}, F_{\mathrm{DLFD}^{(\mathrm{train})}}\right) = \sqrt{\sum_{p=1}^{M/2^s}\sum_{q=1}^{N/2^s}\left(F_{\mathrm{DLFD}_{\mathrm{pq}}^{(\mathrm{test})}}, F_{\mathrm{DLFD}_{\mathrm{pq}}^{(\mathrm{train})}}\right)^2} \tag{4.8}$$

The nearest neighbor classifier is used for identification.

4.3 Results and Discussions

To test the effectiveness of the proposed algorithm for seed identification, we use the image databaseI of the gramineous-grass seeds. In addition to the first batch seed images introduced in Sect. 2.3.1, 5 species of gramineous seeds were supplemented in the seed image database, including Roegneria stricta Keng, Elymus breviaristatus (Keng) Keng f., Leymus secalinus (Georgi) Tzvel., Roegneria ciliaris(Trin.)Nevski, and Roegneria turczaninovii (Drob.) Nevski. 10 seeds were collected for each species, and each of them was captured five times to reduce the effect caused by the variations of focuses, orientations and positions during image acquisition. Therefore, there were 19 classes in the database, and each class harbored 50 image samples taken from 10 seeds. Therefore, there were 19 classes in the database, and each class harbored 50 image samples taken from 10 seeds.

Two groups of experiments were carried out as follows, one was to determine the optimum parameters for division strategy, and the other was to compare the identification performance of different algorithms. All of the experiments were executed on an Intel Core i5-2467M CPU @ 1.60 GHz and 6 GB RAM; the codes were written in MATLAB 7.0.

4.3.1 Local Division Strategy

As we know, the ROIs cropped from original images are of different sizes because of the diversity in geometry size and shape of forage seeds, and auto-focusing without fixed camera equipment in image acquisition. For further local feature extraction, the ROI images should be normalized into a uniform size. Seeing that the outlines of gramineous seeds are mostly elliptical, the ratios of height to width were set from 1:1 to 1:10, with normalized resolutions from 64×64 to 64×640. In addition, the size of partition area is another important factor in local division strategy, which we set 64×64, 32×32, 16×16, 8×8 and 4×4, respectively. By comparing the identification ratios of LFD and DLFD, the optimum combination of normalized size and division area could be found.

Table 4.1 compares the identification accuracies and feature dimensions of LFD and DLFD at different normalized sizes and local partition areas. When the division area is relatively large, the local fractal dimensions of few local blocks cannot well describe the texture of the image completely, yielding poor identification performance. Especially for a 64×64 normalized image with the partition area 64×64, the only feature reveals an identification accuracy of 26.11%. That proves the global fractal dimension cannot represent the whole image effectively. The DLFD is of no meaning because of no difference can be calculated between the only local fractal dimension and the average. The identification accuracies rise with the increase of normalized size and feature dimension, in which the identification performance of DLFD is not as good as that of LFD until the normalized size reaches 64×384. The identification accuracies of LFD and DLFD arrive at 67.58% and 71.58%, respectively, with 6 feature components divided by 64×64 local partitions. The reason mainly lies in that the average of few partitions makes no sense as a reference of texture roughness.

For the local partition area 32×32, the identification accuracy of DLFD is 95.89, 4.84% higher than that of LFD 91.05% within a normalized size 64×640. For the local partition area 16×16, the top identification accuracy of DLFD is always higher than that of LFD. For the local division area 8×8, the identification accuracy of DLFD increases to 98.95% when normalized to size 64×128 with a feature dimension 128, while the identification accuracy of LFD decreases to 98.84%. The enhancement is mainly because DLFD magnify the difference of intrinsic texture by taking the average fractal dimension as the reference, more precisely presenting the roughness of self-similarity texture.

Table 4.1 Comparison of identification accuracies (%) and feature dimensions of LFD and DLFD under different normalized size, local partition area

Normalized size	Algorithms	Local division area				
		64 × 64	32 × 32	16 × 16	8 × 8	4 × 4
64 × 64	LFD	**26.11**	62.00	96.53	98.63	98.84
	DLFD	–	54.84	96.74	98.84	98.84
	Feature dimension	**1**	4	16	64	256
64 × 128	LFD	30.53	**82.11**	98.00	**98.84**	**98.84**
	DLFD	18.00	**85.89**	98.53	**98.95**	**98.84**
	Feature dimension	2	**8**	32	**128**	**512**
64 × 192	LFD	44.42	88.21	97.16	98.84	98.84
	DLFD	27.26	92.21	98.42	98.95	98.63
	Feature dimension	3	12	48	192	768
64 × 256	LFD	53.47	87.89	97.26	98.74	98.84
	DLFD	46.32	92.84	98.74	98.95	98.63
	Feature dimension	4	16	64	256	1024
64 × 320	LFD	63.68	89.58	97.58	98.53	98.74
	DLFD	63.05	94.74	98.42	98.74	98.63
	Feature dimension	5	20	80	320	1280
64 × 384	LFD	**67.58**	90.63	97.47	98.53	98.74
	DLFD	**71.58**	94.74	98.53	98.84	98.63
	Feature dimension	**6**	24	96	384	1536
64 × 448	LFD	70.63	90.63	97.58	98.63	98.63
	DLFD	76.42	95.05	98.53	98.84	98.63
	Feature dimension	7	28	112	448	1792
64 × 512	LFD	72.63	90.21	97.68	98.53	98.63
	DLFD	82.00	95.26	98.53	98.95	98.63
	Feature dimension	8	32	128	512	2048
64 × 576	LFD	74.84	81.20	97.68	98.53	98.53
	DLFD	83.47	95.58	98.53	98.84	98.63
	Feature dimension	9	36	144	576	2304
64 × 640	LFD	**76.32**	**91.05**	97.79	98.53	98.53
	DLFD	**84.63**	**95.89**	98.63	98.95	98.63
	Feature dimension	**10**	**40**	160	640	2560

However, for the local division area 4 × 4, the identification performance of DLFD do not reveal better performance than that of LFD. They are both 98.84% for the normalized 64 × 64 and 64 × 128. With the increase of normalized size, the recognition accuracy of DLFD keep stable, inferior to that of LFD. Small partition area result in more precise description in local texture of LFD, while the differences of LFD and the average (DLFD) of too small partitions magnify specific details, including noise which causes the deterioration of the performance. Moreover, large feature dimensions induced by too many small partitions burden computation and storage, leading to low identification efficiency.

Table 4.1 demonstrates a better overall performance of DLFD as compared to LFD. The top identification accuracy 98.95% of DLFD corresponds to 64 × 128 normalized size, 8 × 8 local division area and 128 feature dimension, selected as the optimum partition strategy in the following experiments. It can be concluded the ratios of length to width do not affect the identification accuracy obviously as observed, suggesting the identification performance is not closely related to the shape contours as we supposed. Comparatively, local block divisions influence the identification performance more obviously; too large local partition areas fail to describe the inner texture variation precisely, while too small partition areas magnify the noise and details at the same time.

4.3.2 Comparison of Different Seed Identification Algorithms

To testify the effectiveness of DLFD, we compare the identification performance of different algorithms, including Fast Fourier Transform FFT_based DLFD, and appearance_based algorithm. The FFT_based DLFD is to convolve FFT with the original image, and then we normalize the imaginary parts to integers of 0 and 255 for a convenient calculation for fractal dimension using DBC. Having divided the FFT transformed image into equal partitions, the DLFD of can be obtained according to the calculation equations of Sect. 4.2. The normalized size, division area, and the feature dimension of FFT_based DLFD were set 64 × 128, 8 × 8 and 128, respectively, correspondence to the DLFD in Sect. 4.3.1.Table 4.2 shows the identification accuracies of different algorithms.

Although FFT is one of the classical tools for image transformation to frequency domain, the identification rate of FFT_based DLFD is 93.16, 5.43% inferior to that of DLFD based on the original image. The deterioration is mainly because the data type of FFT magnitude is float without obvious difference, while the FD obtained by DBC relies highly on the difference of minimum and maximum of gray scales within the local division area.

Table 4.2 Identification accuracies (%) using and the corresponding feature dimension of different algorithms

Algorithms	Identification accuracy	Feature dimension
DLFD	98.59	128
LFD	98.84	128
FFT + DLFD	93.16	128

4.4 Summary

In this paper, we present a seed image based identification algorithm for gramineous grass. In the algorithm, fractal dimensions of all evenly divided local blocks were computed for textual description of the seed capsule. Thus average fractal dimension of all the blocks can be the basis for DLFD feature extraction. By subtracting the individual fractal dimension and the average, the feature vectors were composed of DLFD of all blocks with an emphasis on the contrast of the self-similarity of the seeds.

The algorithm was tested on the gramineous seed base and compared with other classical seed identification approaches. The high accuracies achieved in this task suggested the effectiveness DLFD in intrinsic textual description of seed identification of textual analysis. The novelty lies in the extraction of the difference of local fractal dimension as the textural feature, taking advantage of the local self-similarity in seed images of gramineous grass, which is quite a new application area for pattern recognition. With more and more attention focused on forage identification and grassland monitor for the improvement of living conditions of human beings' ecosystem, computer vision is an essential tool to extend the application to a wider range and a more sophisticated extent.

References

Falconer, K. J. 1990. Fractal geometry: Mathematical foundations and applications. John Wiely and Sons.

Goncalces, W. N., and O. M. Bruno. 2013. Combining fractal and deterministic walkers for texture analysis and classification [J], Pattern Recognition. 46:2953–2968.

Ishak, A. J., A. Hussain, and M. M. Mustafa. 2009. Weed image classification using Gabor wavelet and gradient field distribution[J], Computers and Electronics in Agriculture. 66:53–61.

Mandelbrot, B. 1982. The fractal geometry of nature[D], Freeman.

Pan, X., T. Chen, and Y. Ma. 2016. Seed identification of gramineous grass based on difference of local fractal dimensions[J]. The Open Cybernetics & Systemics Journal. 10, 147–154.

Sarkar, N., and B. B. Chaudhuri. 1994. An efficient differential box-counting approach to compute fractal dimensions of image[J], IEEE Transactions on System, Man, and Cybernetics. 24(1):115–120.

Shanmugavadivu, P., and V. Sivakumar. 2012. Fractal dimensions based texture analysis of Digital Images[J], Procedia Engineering. 38:2981–2986.

Zou, J., Q. Ji, and G. Nagy. 2007. A comparative study of local matching approach for face recognition[J], IEEE Transaction on Image Processing. 16(10):2617–2628.

Chapter 5
Identification of Gramineous Grass Seeds Using Local Similarity Pattern and Linear Discriminant Analysis

In this chapter, we investigated the seed identification for gramineous grass with an approach integrated LSP and LDA to extract the textural features as the input of supervised classification for automatic seed identification of gramineous grass. Unlike some traditional texture feature descriptors, such as HOG and LBP, which are sensitive to noises, LSP is robust to the noise in the real world. In the phase of classifier selection, clustering algorithm based on traditional Euclidean distance considered all the attributes in the clustering have the same effects, and therefore, sometimes it cannot accurately describe the similarity between objects. Linear Discriminant Analysis (LDA) can choose a projection direction to ensure the maximum between-class distance and minimum within-class distance of the samples in the new subspace by adjusting the weight vector components (Fukunaga 2013). So, in this chapter, LSP and LDA were integrated to solve the high similarity of gramineous grass seeds for better identification results (Pan et al. 2017). Figure 5.1 shows the flowchart of the proposed approach.

5.1 Identification Based on LSP and LDA

LSP (local similarity patterns) was proposed by Pourreza in 2011 (Pourreza et al. 2011). It was a kind of rotation invariant operator based on a variety of textural operators with advantages of simple operation, easy understanding, robustness to the variations aroused by grayscale, better identification performance, and etc. Compared with LBP, LSP was insensitive to noise and more powerful in texture analysis. The main difference between LSP and LBP was the selection of threshold. In LBP, the threshold was the grayscale of the center pixel of the neighborhood, while the threshold can be flexibly set with different values in LSP.

The main procedure of LSP was to calculate the absolute differences between the pixels of a 3×3 neighborhood and the center pixel. If the value was greater than a certain SRR, the neighbor pixel was set 0. Otherwise, it would be set 1 (Khunkhett

X. Pan et al., *Computer Vision based Identification and Mosaic of Gramineous Grass Seeds*, https://doi.org/10.1007/978-981-16-3501-4_5

Fig. 5.1 Flowchart of the proposed method

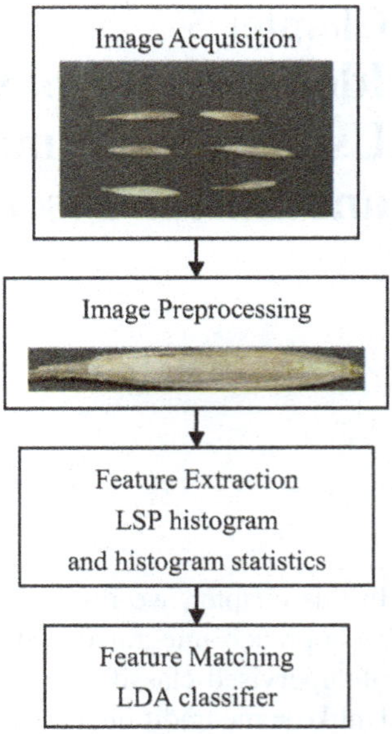

and Remsungnen 2014). Suppose g_c was the center pixel of the neighborhood, and $g_0, g_1, \ldots g_7$ were the pixels of its neighborhood, then the texture T can be converted into binary as follows.

$$T \approx t(s|g_c - g_0| - \mathrm{SRR}), \ldots, s(|g_c - g_7| - \mathrm{SRR}) \tag{5.1}$$

where binary operation was

$$s(x) = \begin{cases} 0, x > 0 \\ 1, x < 0 \end{cases} \tag{5.2}$$

where $x = |g_c - g_0| - \mathrm{SRR}$. Comparatively, LSP is more flexible in the feature selection because of dynamic selection of SRR. When SRR is 0, LSP equals to LBP.

Then the LSP value is the sum of products between the $s(x)$ of all 8-neighborhood pixels and the corresponding weights. Figure 5.2 gives an example when SRR is 10.

Hence, there were 256 LSP values ranged from 0 to 255 altogether. Arrange the binary values of 8-neighborhood pixels clockwise from 8 starting positions, 8 different decimal values would be obtained (Fig. 5.3). The minimum of the 8 values was chosen as the rotation-invariant LSP descriptor of the center pixel. When we selected sampling points within an 8-neighborhood region, there were 36 rotation-invariant LSP values altogether.

In fact, most modes centered on several values, i.e. the histogram was sparse. For a LSP descriptor, the conversion from 0 to 1 or 1 to 0 was called as a jump. If the jump number of a LSP descriptor was no more than 2, it was referred

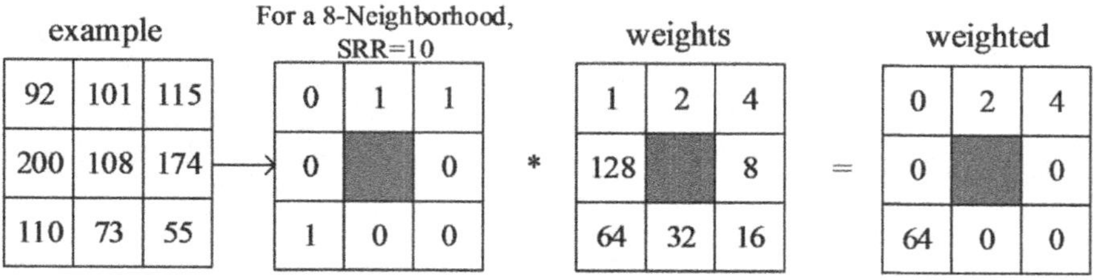

Fig. 5.2 Calculation of LSP value (LSP value: 2 + 4+64 = 70)

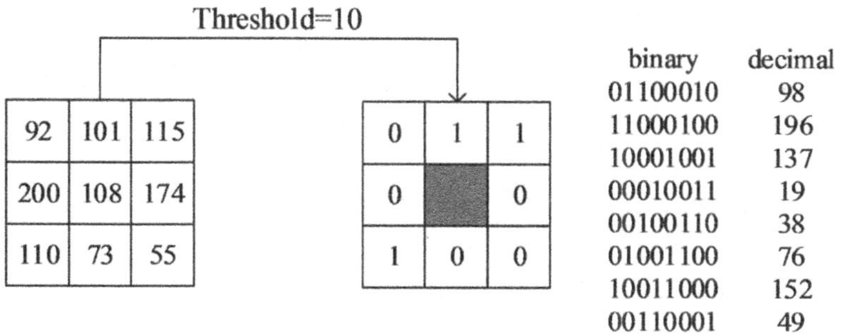

Fig. 5.3 LSP descriptor

to as uniform pattern. For the most LSP descriptors, the jump numbers greater than 2 which were often caused by noises have no statistical meaning. Hence, the numbers of LSP patterns were condensed greatly without losing any information. Meanwhile, the redundant information containing noise was eliminated, along with the desired dimension reduction. When sampled in an 8-neighborhood of 3 × 3 region, 9 LSP descriptors among 36 original ones conformed to "uniform" definition, including 00000000, 00000001, 00010011, 00000111, 01111111, 00011111, 11111111, 01111111, 11111111. Furthermore, the remaining 27 non-uniform descriptors were combined into one descriptor, and hence 10 values were contained in uniform LSP histogram.

When describing the image characteristics, the statistical features of image histogram including mean, standard deviation, smoothness and the third moment can represent the textures effectively. Therefore, 4 image histogram statistical characteristics (Zheng et al. 2010) were concatenated to LSP histogram to form the input the LDA classifiers. The concrete calculation formulas were shown in Table 5.1.

LDA is a classical supervised learning approach to find the optimal combination of features separating two classes with low computational requirements and good classification results. It ensures the projected model in the space with the best separability. Some advanced extensions of LDA have been recently proposed and widely used in many applications of recognition, such as event-related potential (Zhang et al.

Table 5.1 Formulas of statistical features

Feature	Formula
Mean	$\mu = \sum_i p(i)$
Standard deviation	$\sigma = \sqrt{\sum_i (i - \mu)^2 p(i)}$
Smoothness	$1 - 1/(1 + \sigma^2)$
Third moment	$\sum_i (i - \mu)^3 p(i)$

2013a, b, 2014, 2016), electromyography (Song et al. 2007), and etc. They can well solve the problem when insufficient training samples are available. In our approach, the 10 uniform LSP histogram values and 4 histogram statistics, totally 14 features, were imported to LDA classifier for discriminate classification.

5.2 Experimental Results and Discussions

5.2.1 Image Database and Preprocessing

5.2.1.1 Image DatabaseII

For efficient and practical applications, 6 seeds were arranged on a black card when taking photos outdoors as we introduced previously in Sect. 3.3.3.1. In this chapter, we supplemented the seed species. The concrete 12 species and their belonged 5 Genus were listed in Table 5.2. Figure 5.4 gave the original seed image examples of all 12 species in the database.

Table 5.2 Species Names and Genus in the Database

No.	Species	Genus
I	Leymus chinensis(Trii)Tzvel	Leymus
II	Bromus inermis Leyss	Bromus
III*	Elymas sibiricus Linn	Elymus
IV*	Elymus cylindricus (Frmch)Hinda	Elymus
V	Elymus nutans Griseh	Elymus
VI*	Agropyron cristatum var.pectiniforme (Roem.et Schult)H Yang	Agropyron
VII	Agropyron mongolicum Keng	Agropyron
VIII	Agropyron desertorum(Fisch.)Schult	Agropyron
IX*	Roegneria tunczaninovii(Drob.)Nevski var.macrathera Ohwi	Roegneria
X*	Roegneria varia Keng	Roegneria
XI*	Roegneria ciliaris(Trin.)Nevski	Roegneria
XII	Roegneria kamoji Ohwi	Roegneria

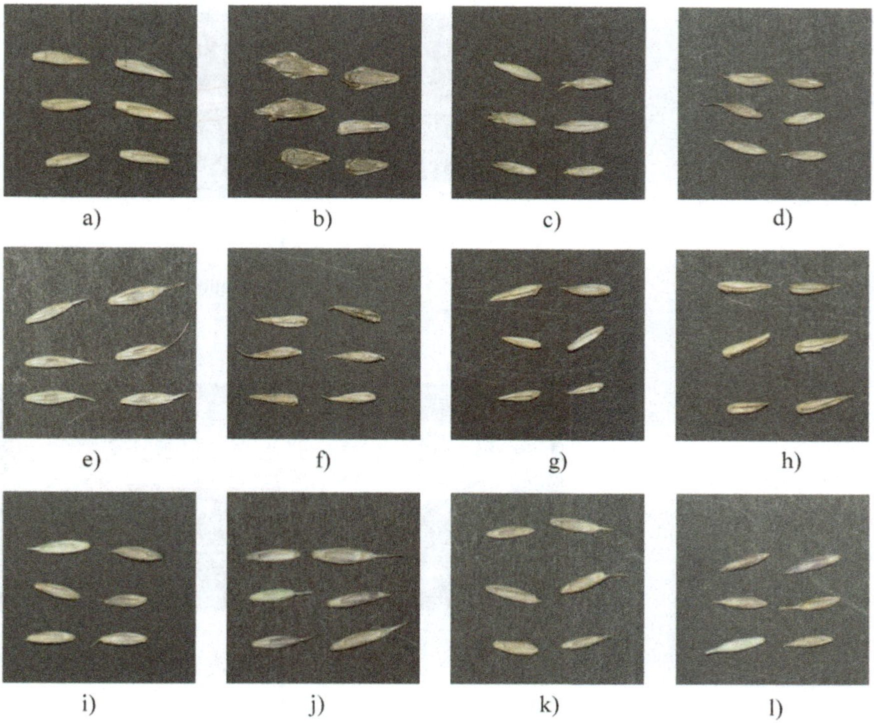

Fig. 5.4 Original images of gramineous grass seeds **a** Leymus chinensis (Trii) Tzvel. **b** Bromus inermis Leyss. **c** Elymas sibiricus Linn. **d** Elymus cylindricus (Frmch)Hinda. **e** Elymus nutans Griseh. **f** Agropyron cristatum var. pectiniforme (Roem.et Schult)H Yang. **g** Agropyron mongolicum Keng. **h** Agropyron desertorum(Fisch.)Schult. **i** Roegneria tunczaninovii (Drob.)Nevski var.macrathera Ohwi. **j** Roegneria varia Keng. **k** Roegneria ciliaris(Trin.) Nevski. **l** Roegneria kamoji Ohwi

5.2.1.2 Image Preprocessing

The main purposes of image preprocessing were to remove the noises that may affect the identification result, and extract region of interest (ROI) from the background. Seen from Fig. 5.4, the outlines of the original seeds in the image directed inconsistently with different tilting angles. Therefore, three steps were involved in image preprocessing as follows.

Firstly, an original color image was converted into the corresponding binary one and the long axis of each seed was detected. Then the image was rotated to keep the long axis horizontal. Thirdly, the sub-images of the seeds were cropped individually from the original image by removing the redundant background. The whole preprocessing procedure was shown in Fig. 5.5. The overall 1080 seed images from 12 species of gramineous grass constructed the image database for experiments. Figure 5.6 gave some examples of the database, where each species has 90 seeds,

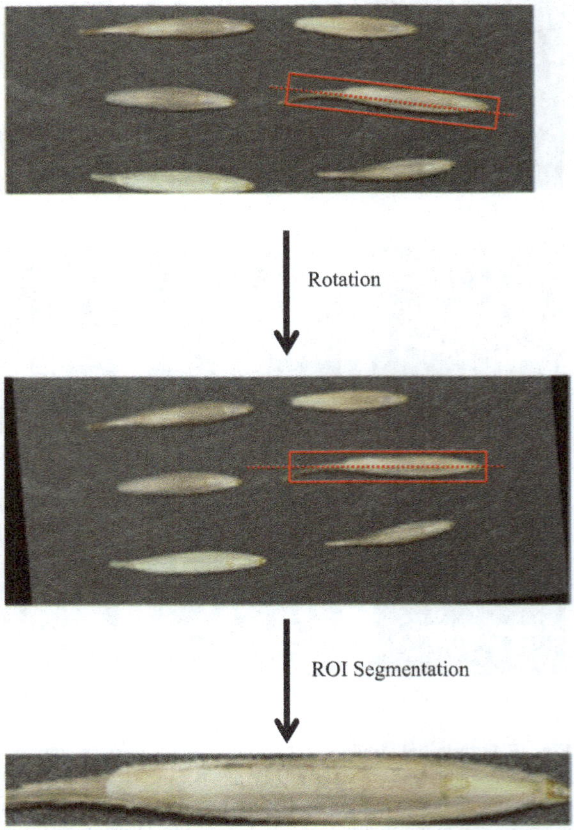

Fig. 5.5 Image preprocessing

and each seed one mere image. All the experiments were executed on an Intel Dual Core i5-3470 CPU @ 1.60 GHz and 4 GB RAM; the codes were written in MATLAB 2011b.

5.2.2 *Identification Experiments*

To compare the effectiveness of the algorithm in distinguishing similar gramineous seeds, two groups of experiments were conducted on different numbers of seed species. In the first group of experiments (Experiment 1), we chose 6 species of gramineous seeds (Roegneria kamoji Ohwi, Bromus inermis Leyss, Elymus nutans Griseh, Agropyron mongolicum Keng, Agropyron desertorum(Fisch.) Schult, Leymus chinensis(Trii)Tzvel) belonged to 5 genus (Roegneria, Bromus, Elymus, Agropyron, Leymus). Only one or two species were contained in each genus, and the textural differences were relatively obvious. In the second group of experiments (Experiment 2), we supplemented other 6 species of gramineous seeds (Roegneria

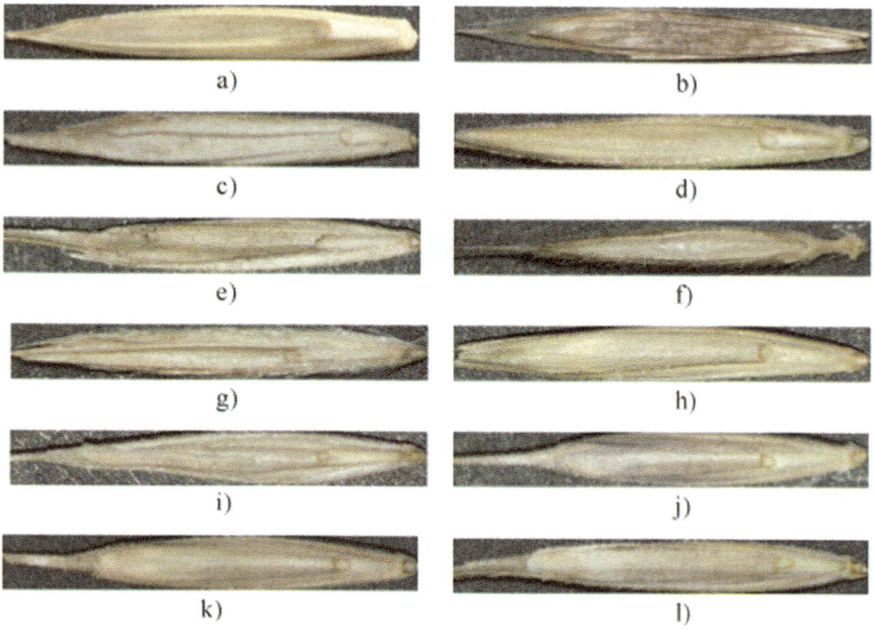

Fig. 5.6 ROI images of 12 species. **a** Leymus chinensis(Trin.) Tzvel. **b** Bromus inermis Leyss. **c** Elymas sibiricus Linn. **d** Elymus cylindricus (Frmch)Hinda. **e** Elymus nutans Griseh. **f** Agropyron cristatum var. pectiniforme (Roem.et Schult)H Yang. **g** Agropyron mongolicum Keng. **h** Agropyron desertorum (Fisch.) Schult. **i** Roegneria tunczaninovii (Drob.) Nevski var. macrathera Ohwi. **j** Roegneria varia Keng. **k** Roegneria ciliaris(Trin.)Nevski. **l** Roegneria kamoji Ohwi

tunczaninovii (Drob.) Nevski var. macrathera Ohwi, Roegneria varia Keng, Roegneria ciliaris (Trin.) Nevski, Elymas sibiricus Linn, Elymus cylindricus (Frmch) Hinda, Agropyron cristatum var. pectiniforme (Roem.et Schult)H Yang) belonged to the 5 above same genus. Thus, more species with very similar texture in the database made the identification task more difficult. The asterisk "*" in Table 5.2 marked the supplemented species in Experiment 2.

For each species, 90 seed images were divided into training set and test set equally, that is either the training set or test set contained 45 images. To avoid the impact of sample selection on the experimental results, cross validation was adopted. By dividing 90 samples of each species into 9 subsets, we selected 5 samples in the same orders from each subset as the training set, and the remaining samples were categorized to test set. For example, when we chose seed images from No. 1 to No. 5 of each subset for training, the remaining seed images from No. 6 to No. 10 composed test set. The average identification result of 126 selections was used as the final results, and the deviation measured the robustness of the identification experiments.

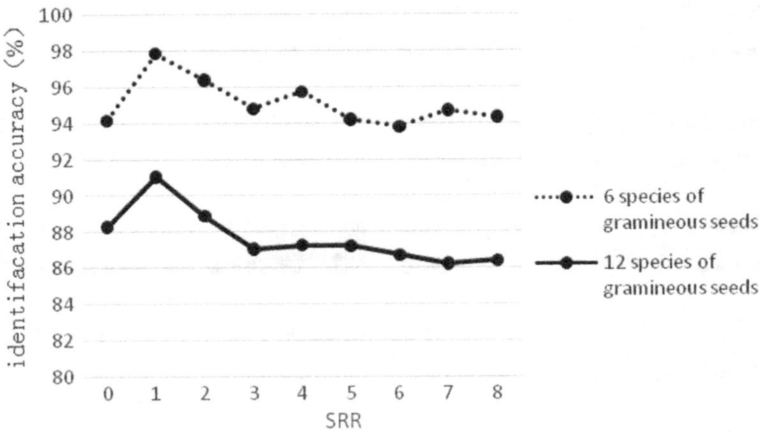

Fig. 5.7 SRR selection

5.2.2.1 Selection of SRR

As we know, different values of SRR would construct various LSP matrices, and accordingly leading to diverse identification results. Figure 5.7 showed the relationship of SRR and identification accuracy when the values of SRR ranged from 0 to 8 in LSP (LDA Classifier). When SRR was 0, LSP equaled to LBP. When SRR was 1, the identification results of 6 seeds species and 12 seeds species were 91.07% and 97.85%, respectively. With the increase of SRR, the identification results declined obviously. Therefore, in the following experiments, SRR was set 1.

5.2.2.2 Comparative Experiments

To testify the performance of the algorithm, we compared 3 different feature extraction approaches (Histograms of Oriented Gradients (HOG), LBP and LSP) and 2 classifiers (Nearest Neighbor Classifier with Eulidean Distance(NNC + ED) and LDA) in the experiments. The sliding steps of HOG (Dalal et al. 2005) were fixed, and the size of image and other parameters were positive correlation. The specified steps were as follows. Firstly, choose a quarter of horizontal and vertical step width L/4 and C/4 as the length of corresponding directions, and fix the step number to 3. Then divide the gradient direction uniformly into 9 bin directions. Let Bsize and Csize represent the sizes of block and cell, then bSize and cSize were $L/2 \times C/2$ and $L/2 \times C/2$, respectively. Table 5.3 listed the identification accuracies of different combinations.

In Experiment 1 conducted on 6 species of seeds, HOG, LBP and LSP did not reveal satisfying identification performances when using NNC + ED classifier. LSP achieved the highest identification accuracy of 80.99%. Comparatively, all the feature extraction algorithms achieved higher accuracies when using LDA classifier. LSP +

Table 5.3 Identification accuracy (%) of different approaches

Feature	Experiment 1 (6 species)				Experiment 2 (12 species)			
	NNC + ED		LDA		NNC + ED		LDA	
	AVG (%)	STDEV (%)	AVG (%)	STDEV (%)	AVG (%)	STDEV (%)	AVG (%)	STDEV (%)
HOG	77.03	0.58	83.19	0.91	46.84	0.97	67.09	0.80
LBP	79.27	0.69	94.18	0.60	57.65	0.86	88.33	0.47
LSP	80.99	0.55	97.85	0.30	60.65	0.70	91.07	0.57

LDA yielded the top identification accuracy of 97.85%. Experiment 1 revealed that LSP can extract the textural feature more precisely as compared with HOG and LBP. Moreover, LDA classifier was more discriminative as compared with traditional NNC + ED.

In Experiment 2 conducted on 12 species of seeds, the identification performances of HOG, LBP and LSP declined obviously when using NNC + ED classifier. The identification accuracy of HOG dropped from 77.03 to 46.84% with a gap of 30.19%. LSP yielded the top identification accuracy of 60.65, 20.43% lower than that of the first group. The decline mainly came from increased 6 kinds of similar seeds, which made the identification more difficult. Comparatively, when using LDA classifier, the three feature extraction approaches achieved much higher accuracies. The identification accuracy of HOG + LDA increased to 67.09, 20.25% higher than that of NNC + ED. LSP + LDA classifier yielded the top identification accuracy of 91.07, 31.09% higher than that of NNC + ED. It can be concluded LDA was more discriminative when identification difficulties increased with more similar species. LSP + LDA achieved more robust identification performance in Experiment 2 with an identification accuracy of 91.07, only 6.78% lower than that of Experiment 1. The standard deviations of the two experiments were lower than 1%, indicating the stability of the overall experiments and the robustness of the algorithm.

It could be observed when the number of seed species was relatively small with less texture similarities, the approaches HOG, LBP and LSP yielded good performance. With the increase of seed species number and identification difficulty level, HOG revealed an obviously decline. The main reason lay in that HOG operated on the local square units, being invariant to geometry and optical deformation appeared on a relatively large region. However, for the similar gramineous grass seeds, more textural detailed were ignored when partitioned to local blocks, leading to a relatively poor identification result. Comparatively, LBP and LSP still worked well with accuracies above 90% because they can detect more details in feature extraction. Moreover, their rotation invariance can deal with the rotations of the biological characteristics and surroundings. The top identification accuracy was achieved by LSP + LDA, indicating LSP was more robust to noise as compared with LBP, and LDA classifier was more discriminative based on category knowledge than NNC + ED.

Table 5.4 Classification rates (%) of HOG + LDA (Experiment 2,12 species)

(a)

Input species no.	Output species no.											
	I	II	III	IV	V	VI	VII	VIII	IX	X	XI	XII
I	72.82	3.03	0.09	3.97	2.24	2.82	–	12.68	–	–	2.35	–
II	8.87	70.74	–	5.38	1.87	0.25	0.07	–	10.51	–	1.31	1.01
III	0.92	0.58	88.77	4.43	2.35	1.23	0.99	0.09	–	–	0.65	–
IV	1.20	1.39	5.96	58.15	7.50	2.10	0.86	–	13.19	5.84	1.78	2.03
V	2.20	3.88	14.11	4.37	64.66	–	1.48	2.96	2.26	4.07	–	–
VI	1.01	2.84	0.02	1.64	0.88	72.40	6.31	6.95	2.19	0.35	4.80	0.62
VII	0.72	1.27	–	1.68	–	4.44	75.45	4.60	2.31	–	–	9.52
VIII	7.46	–	–	–	0.42	6.03	4.27	75.77	1.75	1.90	2.40	–
IX	–	6.56	2.33	8.04	5.52	7.18	–	2.59	27.60	9.12	16.97	14.09
X	0.71	0.07	–	2.38	–	1.13	–	7.07	2.40	53.33	24.23	8.68
XI	4.94	6.40	–	0.78	0.49	4.71	–	0.19	5.29	3.67	67.44	6.08
XII	1.78	1.92	–	1.39	1.23	–	1.09	1.53	1.29	5.70	6.10	77.95

(continued)

Table 5.4 (continued)

(a)

		Output species no.											
(b)													
Input species no.		I	II	III	IV	V	VI	VII	VIII	IX	X	XI	XII
	I	93.46	–	–	–	–	–	–	6.54	–	–	–	–
	II	–	64.69	13.54	21.76	–	–	–	–	–	–	–	–
	III	–		100.00	–	–	–	–	–	–	–	–	–
	IV	–	–	–	98.15	1.85	–	–	–	–	–	–	–
	V	–	–	–	2.86	95.52	0.35	1.27	–	–	–	–	–
	VI	–	–	–	–	1.47	88.64	6.82	3.07	–	–	–	–
	VII	–	–	–	–	–	7.44	89.35	3.21	–	–	–	–
	VIII	2.35	–	–	–	–	4.81	5.49	81.04	6.31	–	–	–
	IX	–	–	–	–	–	1.59	–	5.59	90.21	2.61	–	–
	X	–	–	–	–	–	–	–	–	–	79.31	12.52	8.17
	XI	–	–	–	–	–	–	–	–	2.52	4.07	83.26	10.15
	XII	–	–	–	–	–	–	–	–	–	2.48	1.21	96.31

(continued)

Table 5.4 (continued)

(a)

(c)

Input species no.	Output species no.											
	I	II	III	IV	V	VI	VII	VIII	IX	X	XI	XII
I	94.76	–	–	–	–	–	–	5.24	–	–	–	–
II	–	70.72	13.95	15.33	–	–	–	–	–	–	–	–
III	–	–	100.00	–	–	–	–	–	–	–	–	–
IV	–	–	–	100.00	–	–	–	–	–	–	–	–
V	–	–	–	2.57	97.43	–	–	–	–	–	–	–
VI	–	–	–	–	–	94.25	2.77	2.98	–	–	–	–
VII	–	0.07	–	–	–	2.22	97.09	0.62	–	–	–	–
VIII	2.40	–	–	–	–	7.76	8.22	81.62	–	–	–	–
IX	–	–	–	–	–	–	–	–	100.00	–	–	–
X	–	–	–	–	–	–	–	–	–	80.67	13.40	5.93
XI	–	–	–	–	–	–	–	–	1.27	5.22	84.92	8.59
XII	–	–	–	–	–	–	–	–	–	2.22	0.69	97.09

To investigate on the overall decline of the Experiment 2, we listed the average category details of 126 cross validation experiments using HOG + LDA, LBP + LDA and LSP + LDA in Table 5.4. The rows and columns corresponded to the number of input and output species, respectively. For example in Table 5.4a, 72.82% seed images were identified correctly to No. I species among 45 samples in the test set, and the remaining samples were mistakenly identified to other species. As can be seen, HOG can not well distinguish the seed images whether they are in the same genus. In LBP and LSP, most of the mistakenly classified seeds were of the same genus with very similar textures. In Table 5.4c, among the 45 test samples of No. I species, 94.76% seeds were correctly classified to No. I species, only 5.24% seeds were mistakenly identified to No. XI species, showing that LSP + LDA classifier was more capable of describing similar textures.

5.3 Summary

In this chapter, we presented a seed identification algorithm using LSP and LDA for gramineous grass. The seeds of the gramineous grass contained very similar in texture, color, shape and size, which made the identification task difficult. Moreover, the circumstances of image acquisition outdoors gave rise to noises, variations of rotation and scales to the seed images. Therefore, LSP + LDA can well solve the problems, in which the former can extract more specific textures robust to noise and rotation variance, and the latter was more discriminative with classification information. The algorithm was tested on a gramineous seed base composed of 12 similar species, yielding an identification accuracy of 91.07%. The novelty existed in the utilization of LSP + LDA in a highly similar gramineous seed identification task. Machine vision was proved to be an essential tool to enhance the automated level on forage identification and grassland monitor.

References

Dalal, N., and B. Triggs. 2005. Histograms of oriented gradients for human detection[C], in 2005 IEEE Computer Society Conference on Computer Vision and Pattern Recognition, 886–893.

Fukunaga, K. 2013. Introduction to statistical pattern recognition[M]. Academic press, 60:2133–2143.

Khunkhett, S., and T. Remsungnen. 2014. Non-destructive identification of pure breeding Rice seed using digital image analysis[C], in Information and Communication Technology, Electronic and Electrical Engineering (JICTEE), 1–4.

Pan, X., T. Chen, Y. B. Ma, and Z. Yu. 2017. Seed Identification of Gramineous Grass Using Local Similarity Pattern and Linear Discriminant Analysis[J].The Open Cybernetics &Systemics Journal, 11, 108–118.

Pourreza, H. R., M. Masoudifar, and M. ManafZade. 2011. LSP: Local similarity pattern, a new approach for rotation invariant noisy texture analysis[C], in 2011 18th IEEE International Conference on Image Processing (ICIP), 2011, 837–840.

Song, Y. C., Y. Y. Zhang, H. D. Meng. 2007. Research based on euclid distance with weights of clustering method[J]. Jisuanji Gongcheng yu Yingyong(Computer Engineering and Applications), 43:179–180. (in Chinese).

Zhang, H., Y. Zhao, F. Yao, L. Xu, and G. Li. 2013a. An adaptation strategy of using LDA classifier for EMG pattern recognition[C]. in Engineering in Medicine and Biology Society (EMBC), 35th Annual International Conference of the IEEE, 4267–4270.

Zhang, Y., G. Zhou, Q. Zhao, J. Jin, X. Wang, and A. Cichocki. 2013b. Spatial-temporal discriminant analysis for ERP-based brain-computer interface[J], IEEE Transactions on Neural Systems and Rehabilitation Engineering, 21:233–243.

Zhang, Y., G. Zhou, J. Jin, Q. Zhao, X. Wang, and A. Cichocki. 2014. Aggregation of sparse linear discriminant analyses for event-related potential classification in brain-computer interface. International Journal of Neural Systems, 24(01), 767–164.

Zhang, Y., G. Zhou, J. Jin, Q. Zhao, and A. Cichocki. 2016. Sparse bayesian classification of EEG for brain–computer interface [J]. IEEE Transactions on Neural Networks and Learning Systems, 27(11):1–1.

Zheng, Z., Y. Wang, and G. Xue. 2010. Digital Image Processing and Machine Vision - Visual C++ and Matlab Implementation[M], POSTS & TELECOM Press. (in Chinese).

Chapter 6
Identification of Gramineous Grass Seeds Using Local Similarity Pattern and Gray Level Co-occurrence Matrix

In this chapter, we continue to focus on the on the rotation invariance of Uniform LSP in the identification of Gramineous Grass Seeds. Fusing with GLCM, the texture characteristics of grass seeds to identify the seeds whose color, shape, sizeand other characteristics are very similar can be extracted. So, in this chapter, LSP and GLCM were integrated to solve the high similarity of gramineous grass seeds for better identification results (Chen et al. 2019).

6.1 Fusion of LSP and GLCM in Feature Extraction

6.1.1 Gray Level Co-occurrence Matrix

Gray level co-occurrence matrix was proposed by Haraklick in 1979 (Haralick 1979), which is one of the early algorithms for describing global texture features of images. The gray level co-occurrence matrix, as a second-order statistic, represents the joint distribution of the spatial position relations of two pixels. The calculation approach of GLCM is as follows.

(1) Take any pixel (x, y) in the image and the other point $(x + a, y + b)$ with a distance of $d = \sqrt{x^2 + y^2}$ and an angle of θ to form a pair of pixels expressed in gray value (i, j).
(2) Move the point (x, y) over the whole image and obtain various (i, j) values.
(3) Count the frequency $P\ (i, j, d)$ of each pair of pixels. Assuming the gray level of the image is L, the square matrix $[P\ (i, j, d, \theta)]L \times L$ is the gray level co-occurrence matrix.

It can be inferred from the above steps that multiple gray level co-occurrence matrices can be derived from various distances D and angles. Usually, the values of angle can be four directions: 0°, 45°, 90°, and 135°.

X. Pan et al., *Computer Vision based Identification and Mosaic of Gramineous Grass Seeds*, https://doi.org/10.1007/978-981-16-3501-4_6

Unlike the other texture features, GLCM is not directly used for subsequent classification after extracting features; instead it carries out some statistical calculations on the basis of matrices to obtain statistics as the feature of texture recognition. The original gray level co-occurrence matrix has only 14 statistics. Subsequently, it is found that only energy, entropy, contrast and correlation are irrelevant. These 4 statistics are not only available for calculation, but also works well in classification (Ulaby et al. 1986).

(1) Energy, also known as second-order moment. The function is to measure the consistency of image texture, which can reflect the uniformity of gray distribution and the thickness of texture. The more uniformed gray scale distribution and fine texture mean more energy. On the contrary, more concentrated gray distribution and rough texture means less energy. The formula is as follows:

$$S_E = \sum_{i=0}^{L-1}\sum_{j=0}^{L-1}\{P(i, j|d, \theta)\}^2 \tag{6.1}$$

(2) Contrast. It is a statistical measure of clarity and depth of grooves in texture reflecting the extent of texture difference. The deeper texture groove has a greater contrast degree as compared with the shallower texture groove. From the perspective of human vision, the higher contrast means the clearer the visual effects. The formula is as follows:

$$S_{Con} = \sum_{i=0}^{L-1}\sum_{j=0}^{L-1}(i - j)^2 \mathrm{P}(i, j|d, \theta) \tag{6.2}$$

(3) Correlation. It is a statistical measure of the local gray correlation of image, reflecting the similarity of elements in GLCM on rows or columns. When the values of elements in the matrix are uniform and equal, the value of correlation is large. On the contrary, the correlation value becomes smaller when the values of elements in the matrix differ greatly. The definitions of μ_1, μ_2, σ_1 and σ_2 are as follows:

$$\mu_1 = \sum_{i=0}^{L-1} i \sum_{j=0}^{L-1} \mathrm{P}(i, j|d, \theta) \tag{6.3}$$

$$\mu_2 = \sum_{j=0}^{L-1} j \sum_{i=0}^{L-1} \mathrm{P}(i, j|d, \theta) \tag{6.4}$$

$$\sigma_1^2 = \sum_{i=0}^{L-1}(i - \mu_1)^2 \sum_{j=0}^{L-1} \mathrm{P}(i, j|d, \theta) \tag{6.5}$$

$$\sigma_2^2 = \sum_{j=0}^{L-1} (j - \mu_2)^2 \sum_{i=0}^{L-1} P(i, j|d, \theta) \tag{6.6}$$

Then correlation is defined as follows:

$$S_{Cor} = \frac{\sum_{i=0}^{L-1} \sum_{j=o}^{L-1} ijP(i, j|d, \theta) - \mu_1\mu_2}{\sigma_1\sigma_2} \tag{6.7}$$

(4) Entropy. It is a statistical measure of the random characteristics of the histograms. The greater the entropy means more randomness, which contains the more information. The formula for calculating entropy is as follows:

$$e = -\sum_{i=0}^{L-1} P(z_i) \log_2 P(z_i) \tag{6.8}$$

6.1.2 Fusion of LSP and GLCM for feature extraction

The fusion of LSP and GLCM can integrate the local and global information of the images. Compared with each single algorithm, the fusion can better represent the texture for an improved identification result.

The basic procedure of the algorithm is listed as following:

(1) The LSP descriptors of all the pixels are calculated by the Uniform $LSP_{8,1}^{ri}$ operator, instead of histogram statistics.
(2) Calculate the grayscale co-occurrence matrix of LSP image processed in Step 1 with GLCM algorithm, where the instance $d = 1$, and orientation $\theta = 0°$, 45°, 90°, 135°
(3) Calculate the 4 statistic parameters of each grayscale co-occurrence matrix, and combine all the statistic values as the input of LDA classifier for identification.

6.2 Experimental Results and Discussions

In this study, the traditional LSP and GLCM methods are used as comparative experiments, followed by LDA classifier for feature matching.The feature dimension of LSP is 14, including 9 equivalent LSPs, 1 non-equivalent LSP and 4 statistical features (mean, standard deviation, smoothness and third-order moments). The feature dimension of GLCM is 16, which is obtained by concating four statistics (energy, contrast, correlation and entropy) from the co-occurrence matrix in 4 directions (0°, 45°, 90° and 135°).

In order to explore the scalability of the algorithm, the image database data is divided into two groups for experiments separately. In experiment 1, 6 kinds of forage seeds (Euphorbia, Bromus inermis, Elymus nutans, Agropyron mongolica, Agropyron deserticola and Leymus chinensis) were selected from 5 different genera (Euphorbia, Bromus, Elymus, Agropyron and Leymus). Their texture differences are relatively obvious. In experiment 2, 6 species of grass seeds (Eupatorium sibiricum, Eupatorium variabilis, Eupatorium villosum, Elymus terminalis and Agropyron glabra) with the same genus or similar texture as those in Experiment 1 were added to enhance the similarity among the samples for more practical applications.

In the experiments, each kind of seed has 90 pictures, 45 of which are selected as training set, and the remaining pictures are used as test set. In order to avoid the impact of sample selection, cross validation is used to ensure the accuracy of the results when selecting training set and data set. 90 samples were divided into nine on average. 5 samples were selected as training samples in the same order in each 10 samples. The remaining 5 samples were used as test samples (e.g. 1–5 as training samples, 6–10 as test samples). There are 252 kinds of methods to select, and the average and standard deviation are taken as the experimental results.

In Table 6.1, the recognition performance of LSP, GLCM and their fusion algorithms are compared. It can be seen that GLCM has the lowest recognition rate (average accuracy) and the largest standard deviation in experiment 1, which contains 6 kinds of seeds. The recognition rate of LSP and GLCM fusion algorithm is close to that of single LSP algorithm, only 0.79% difference, and the standard deviation between them is low. In experiment 2, the recognition rate of GLCM decreased significantly and remained the lowest when the seed types were increased to 12 categories. It can be seen that the GLCM algorithm based on global texture is not effective in identifying gramineous forage seeds, and the results are unstable. The recognition rate of LSP fusion GLCM algorithm is only reduced by 3.87%, while that of LSP algorithm is reduced by 6.78%. This shows that the algorithm is less affected by similar kinds of images and has stronger adaptability than LSP algorithm.

It can be observed from Table 6.1 that the recognition rate of each algorithm will decrease when the number of gramineous forage seed categories increases. In order to explore the reasons for the overall decline of recognition rate, the classification of LSP, LSP and GLCM fusion algorithm are listed in Tables 6.2 and 6.3, respectively, in which LDA is used as classifier. In each table, rows correspond to the category of input seeds to be identified, and columns represent the output category after classification. For example, in Table 6.2, Among the 45 seeds of the first category in the test set, 94.76% of the test samples were correctly identified (the values corresponding to the first row and the first column), 5.24% of the test samples were misclassified into the

Table 6.1 Comparison of identification performance of different algorithms

	Exp.1 (6 species)		Exp.2 (12 speices)	
Algorithms	Average accuracy (%)	STDEV	Average accuracy (%)	STDEV
LSP	97.85	0.30	91.07	0.57
GLCM	60.04	1.38	35.09	0.8
LSP + GLCM	98.64	0.22	94.77	0.4

Table 6.2 Seed classification results of 12 species based on LSP (correct classification rate)

Output species no.												
Input species no.	1	2	3	4	5	6	7	8	9	10	11	12
1	94.76	–	–	–	–	–	–	5.24	–	–	–	–
2	–	70.72	13.95	15.33	–	–	–	–	–	–	–	–
3	–	–	100.0	–	–	–	–	–	–	–	–	–
4	–	–	–	100.0	–	–	–	–	–	–	–	–
5	–	–	–	2.57	97.43	–	–	–	–	–	–	–
6	–	–	–	–	–	94.25	2.77	2.98	–	–	–	–
7	–	0.07	–	–	–	2.22	97.09	0.62	–	–	–	–
8	2.40	–	–	–	–	7.76	8.22	81.62	–	–	–	–
9	–	–	–	–	–	–	–	–	100.0	–	–	–
10	–	–	–	–	–	–	–	–	–	80.67	13.40	5.93
11	–	–	–	–	–	–	–	–	1.27	5.22	84.92	8.59
12	–	–	–	–	–	–	–	–	–	2.22	0.69	97.09

Table 6.3 Seed classification results of 12 species based on the fusion of LSP and GLCM (correct classification rate)

Output species no.												
Input species no.	1	2	3	4	5	6	7	8	9	10	11	12
1	97.78	2.22	–	–	–	–	–	–	–	–	–	–
2	–	100.0	–	–	–	–	–	–	–	–	–	–
3	–	10.66	86.74	2.60	–	–	–	–	–	–	–	–
4	–	–	–	95.56	4.44	–	–	–	–	–	–	–
5	–	–	–	2.21	91.96	5.83	–	–	–	–	–	–
6	–	–	–	–	0.27	99.73	–	–	–	–	–	–
7	–	–	–	–	–	–	93.33	6.67	–	–	–	–
8	–	–	–	–	–	–	1.08	92.25	6.67	–	–	–
9	–	–	–	–	–	–	–	1.38	98.62	–	–	–
10	–	–	–	–	–	–	–	–	3.25	83.95	12.80	–
11	–	–	–	–	–	–	–	–	–	–	99.55	0.45
12	–	–	–	–	–	–	–	–	–	–	2.20	97.80

eighth category (the values corresponding to the first row and the ninth column), and so on.

As shown in Table 6.3, it can be found that most of the misclassified seeds are identified mistakenly into other seed species of the same genus. Basically, it will not be recognized to other seed classes of different genera, which proves that the algorithm is robust in identify seed identification of different genera. At the same time, when the images of similar species from the same gramineous family are added, the recognition rate of the algorithm decreases within a minimum scope, which is better than LSP or GLCM for identical gramineous grass seeds.

References

Chen, T., X. Pan, Y. B. Ma, and W. H. Yan. 2019. Seed feature extraction algorithm of gramineous grass based on the fusion of LSP and GLCM [J], Journal of China Agriculutral University, 24(7):138–145. (in Chinese).

Haralick, R. M. Statistical and structural approaches to texture [J]. Proceedings of the IEEE, 1979, 67(5):786–804.

Ulaby, F. T., F. Kouyate, B. Brisco, and T. H. Williams. 1986. Textural information in SAR images [J]. IEEE Transactions on Geoscience and Remote Sensing, 24(2):235–245.

Chapter 7
Microscopic Image Mosaic of Gramineous Grass Seeds

7.1 Image Mosaic

Image mosaic is originated from human photography knowledge. When the camera's field is smaller than that of human, people will naturally consider splicing multiple photos into a large one for a broader field. However, due to technical and cost constraints, e.g. hardware equipment, it is difficult to quickly generate panoramas. Image mosaic technology can mosaic a group of overlapping image sequences (acquired by common equipment) into a large seamless high-resolution image, which makes the image acquisition equipment more economical and common, and the acquisition of wide and high-resolution images more simpler. In recent years, with the development of computer performance and image processing technology, image mosaic technology based on automatic registration and fusion has become a hot topic in computer vision research.

Image mosaic technology is a seamless and high-definition image processing technology, which combines a group of image sequences with clear correlation among them through intelligent matching calculation and spatial matching alignment. The main process of image mosaic technology is to pre-process the mosaic image, then complete the image registration, and finally get the panorama image by image fusion (Fig. 7.1).

The main purpose of image preprocessing is to eliminate irrelevant information in images, restore useful real information, enhance the detect ability of relevant information and simplify data to the greatest extent, so as to improve the reliability of feature extraction, matching and fusion.

Image registration operation is the process of transforming images from the same scene (or object) to the same coordinate system through the same or different imaging devices at different times or perspectives. Commonly, the registration methods are mainly divided into three categories: based on image gray information, phase correlation and features.

The purpose of image fusion is to fuse complementary information or salient information into one image, extract useful information from two images, remove

X. Pan et al., *Computer Vision based Identification and Mosaic of Gramineous Grass Seeds*, https://doi.org/10.1007/978-981-16-3501-4_7

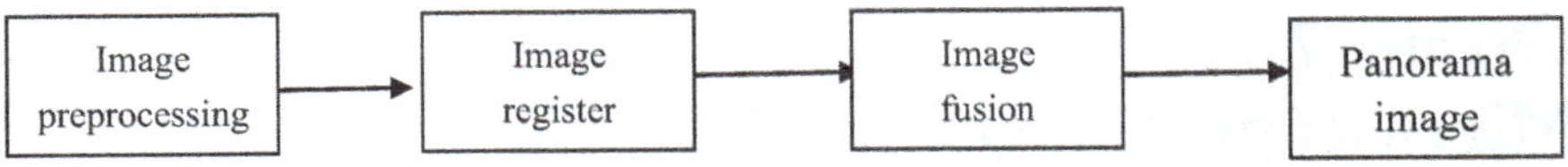

Fig. 7.1 Flowchart of image mosaic

redundant information between images, and finally form a complete image. The commonly used image fusion methods are the average method, the gradual-in–out method and the median filtering fusion method.

Usually, the initial matching contains some wrong matching points. Fischler and Bolles (1987) proposed a simple and highly stable method to remove mismatches, also known as Random Sample Consensus (RANSAC). The experimental results of RANSAC are better than those of other algorithms (Wendy et al.2009). In this paper, RANSAC algorithm is used for error matching. In the fusion stage, the gradual extrusion method is used to smooth the seams between mosaic images, but the effect is not obvious. Moreover, the image processing with illumination changes is not perfect. Therefore, in order to overcome the degradation of image quality caused by illumination changes, Gamma correction method is used to reduce its impact, so the mosaic image is more natural. Finally, the complete image is richer than the single image (Ning and Pan 2013).

7.2 Materials and Methods

7.2.1 Experimental Materials

The acquisition equipment for micro-image of gramineous seed are Motic BA200, the micro lens Motic EF Plan 4 * 1.0 and Camera Motcam 300, as shown in Fig. 7.2.

Fig. 7.2 Image acquisition systems for seed microimage

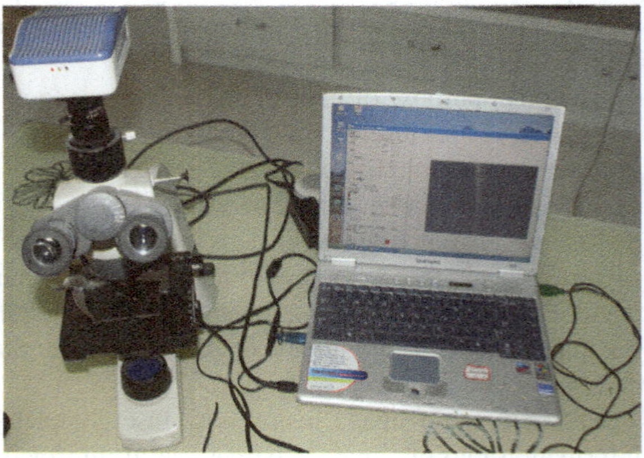

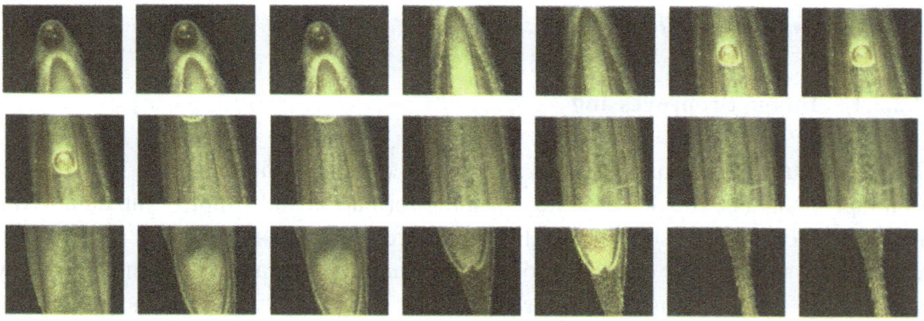

Fig. 7.3 The ventral microimages of Roegneria varia Keng Seed

When collecting images, seeds are placed on black card, and the paper and seeds are placed flat under the objective microscope. Because of the limited vision field of the microscope, the whole seed image cannot be displayed in one image. Therefore, only part of the image can be captured in each photograph, and the multiple micro-images are the components of entire seed image.

Microscopic image of front side of Roegneria varia Keng seed is shown in Fig. 7.3. The resolution, color depth, and the size of the image are 512 × 384, 24 bits, and 577 KB, respectively. The 21 images in Fig. 7.3 (ranking order: from left to right, from top to bottom) are the component micro-images of the entire seed. In the subsequent study, the local seed images or integrated seed images can be directly used to extract features for automatic identification. The corresponding macro photograph of the seed is shown in Fig. 7.4. The acquisition device is SONY-350 and the macro lens MACRO2.8/100.

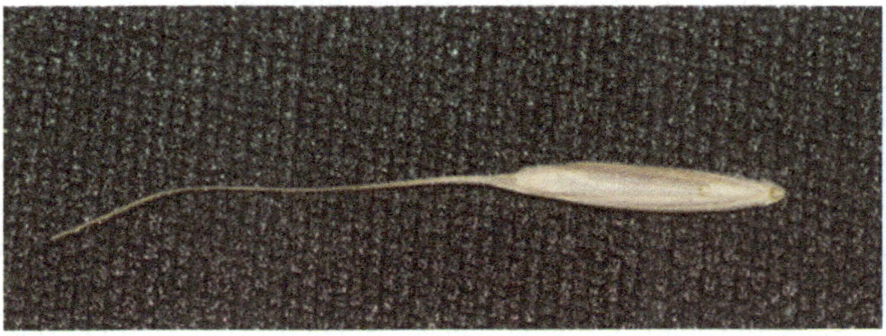

Fig. 7.4 The ventral micro-distance image of Roegneria varia Keng Seed

7.2.2 Mosaic Method

7.2.2.1 Image Preprocessing

The main purpose of image preprocessing is to eliminate the irrelevant information in images, enhance the detect ability of the information and simplify the data to the greatest extent, and ensure the quality of image registration (Wang et al. 2008).In view of the single color of microscopic image and the degradation of image quality affected by illumination, in this algorithm, the preprocessing steps include image graying, removing noises and filtering, histogram averaging and Gamma correction.

Grayscale image processing is the process of transforming color image into gray image. Seed skin color is comparatively similar and has no special distinguished feature. Moreover, color images occupy a larger storage space with a slower processing speed, so the conversion into grayscales can improve the computing speed.

Image noise is the random interference occurred in image acquisition, transmission or processing. So the quality of image is reduced. Image noise is generally divided into two kinds: salt and pepper noise and random noise. These two kinds of noises are contained in the acquired micro-image. Mean filter is used to denoise the image and reduce the noise interference.

Gray level normalization is to adjust the gray level of images to be stitched to the appropriate gray level range, so that it is easy to find matching points. The effect of stitching is more consistent. The histogram equalization method is used to make the gray scale distribution of images more uniform. The gray histogram of the original image can be changed from a relatively centralized gray level interval to a uniform distribution in the whole gray level range (Hou 2007), so as to achieve the goal of less illumination effects on the image.

Because the color images are disturbed by illumination change, camera CCD noise, camera angle change and other factors in the acquisition process, it is necessary to correct the color information of each sub-image. Thus the panoramic image with the same color can be obtained. Therefore, an adaptive Gamma correction method is needed to correct the microscopic image of forage seeds. Gamma correction method has a strong advantage in overcoming the influence of illumination. After analyzing the collected image, the non-linear relationship between Gamma value and pixel value is established, so that the appropriate correction can be made according to the specific value of each pixel in the image (Ma et al. 2010). After correction, the illumination distribution is more uniform and the texture feature is prominent, which is suitable for feature extraction.

7.2.2.2 Image Registration

Image registration operation is the process of transforming images from the same scene (or object) to the same coordinate system through the same or different imaging devices at different times or perspectives (Song 2010). Commonly, the registration methods are mainly divided into three categories according to the based information: image gray information, phase correlation and features. The image registration based on features is adopted in the research. The registration process usually needs to extract feature points from different images and find matching feature points via similarity measurement. Then the transformation parameters of image spatial coordinates, namely the transformation matrix, are obtained according to the pair of feature points. Finally, coordinate transformation matrix is used for image registration.

In order to overcome the noise interference on the view angle and scale caused by camera motion, Scale Invariant Feature Transform (SIFT) algorithm based on multi-scale space theory is used to extract and match the feature points of the image sequence (Lowe 2004). Based on the scale space, the algorithm can match two images invariantly in the case of translation, rotation, scale change and illumination change. To a certain extent, it also has a relatively stable feature matching ability for visual angle change and radiation change, and can extract clear invariant features from different views of the same object or scene. Mikolajczyk and Schmid (2005) pointed out that SIFT algorithm has robustness after assignment operation and comparison of different detection operators and local operators.

The generation of SIFT feature vectors of an image mainly includes four steps: scale space extremum detection, key point location and scale determination, key point direction determination and feature vector generation. The main task of scale space extremum detection is to establish image scale space (or Gauss pyramid) based on scale space theory and detect extremum points. Let $G(x, y, \sigma)$ be a Gaussian kernel function, then the scale space $L(x, y, \sigma)$ of image I (x, y) at different scales is represented as:

$$L(x, y, \sigma) = G(x, y, \sigma) * I(x, y) \tag{7.1}$$

where,

$$G(x, y, \sigma) = \frac{1}{2\pi\sigma^2} e^{-(x^2+y^2)/2\sigma^2} \tag{7.2}$$

is the Gauss kernel function, which is proven to be the only linear kernel function. * is the convolution operator, σ is the scale-space factor, k is the multiplicative factor.

In order to detect stable feature points efficiently in scale space, Low uses the differential Gauss DoG extremum in scale space as the judgment basis. It can be defined as follows:

$$\begin{aligned} D(x, y, \sigma) =& (G(x, y, k\sigma) - G(x, y, \sigma)) * I(x, y) \\ =& L(x, y, k\sigma) - L(x, y, \sigma) \end{aligned} \tag{7.3}$$

In practical scale invariant for feature point extraction, image pyramid is introduced into scale space in SIFT algorithm.

After the pyramid construction, DoG local extremum is detected. In order to find the extreme points in scale space, each pixel in the middle layer of DoG scale space needs to be compared with a total of 26 pixels of 8 pixels in the surrounding area of the same scale and 9 * 2 pixels in the surrounding area corresponding to the location of the adjacent scale. Only when DoG value of the monitored points, also known as the cross-marked pixel in the graph, is greater than or less than the 26 pixel, they are determined as the extreme point and saved for subsequent calculation.

Having obtained the set of SIFT candidate feature points of the original image, stable points should be selected as SIFT feature points of the image. Because the points with low contrast in candidate feature points are sensitive to noise and the points on the edge are difficult to be located accurately. In order to ensure the stability of SIFT feature points, low contrast points and unstable edge response points must be eliminated.

(1) Descriptor of SIFT features. After an image location, scale and orientation hav e been assigned to each key point, the next step is to compute a descriptor for the local image region (Lowe 2004). Firstly, rotate the coordinate to the orientation of the feature point in order to insure the rotation invariant. Secondly, to describe the feature points, its adjacent region of 8 × 8 pixels are usually proposed. This region could be evenly divided into four 4 × 4 sub-regions. After calculating the gradient histogram of 8 directions for each sub-region and make the vectors of 8 directions for each point in an orderly sort, thus a feature vector of 4 × 4 × 8 equal to 128-dimensions is constituted.
(2) Feature matching. When the SIFT feature vectors of two images have been generated, the best candidate match for each key point in the reference image is obtained by identifying its first nearest neighbor and second nearest neighbor with Euclidean distance in all the descriptors of sensed image. If the distance between key point and its first nearest neighbor is d1 and the value between key point and its second nearest neighbor is d2, the ratio between d1 and d2 is less than r, the key point with first nearest neighbor is considered as the best match point. After the above steps, the initial one-to-one matching between key points of the reference image and the sensed image is obtained, that is, $Q = \{q_i\}$ and $Q' = \{q_i'\}$, i = 1, 2, …, N (where q_i matches q_i').
(3) Removing Incorrect Matches. If some incorrect correspondences occur, this might lead to wrong results. Random Sample Consensus (RANSAC) algorithm is used to purifying the matching set, while catching the optimal perspective transformation matrix H. The RANSAC algorithm has the following steps (Lowe 2004):
(1) Strike a maximum sampling frequency N based on probability, repeat sampling N times randomly;
(2) Select four pairs of matching points randomly, in which any of three are not collinear, then calculate the transformation matrix H;
(3) Calculate the distance between each matching point after they have been transformed to the corresponding matching point;

(4) Calculate the number of inner points whose distance is less than the threshold value, screen interior point out from dataset which contains many outer points and correctly estimate model parameters;
(5) Compute the optimal perspective transformation matrix H by using the matching points which have eliminated false matches.

Among the existing image mosaic methods, the Scale Invariant Feature Transform (SIFT) algorithm based on multi-scale space theory is widely used, because the SIFT features can effectively maintain invariance to the rotation, scaling and even affine transformation; moreover, maintain a certain degree of stability and adaptability (Feng et al. 2009; Fischler and Bolles 1981). SIFT algorithm was originally proposed in 1999 by David G. Low, and further improved in 2004 by the same author. The main purpose of SIFT is to extract distinctive invariant features from images between different views of an object or scene. Mikolajczyk pointed out SIFT is the robustness and the distinctive character after evaluating and comparing the various widely used detectors and local descriptors. Generally, the initial matching always contains some incorrect matches. Fischler and Bolles presented a simple and highly robust method named Random Sample Consensus (RANSAC) to remove incorrect matches. Compared to other methods, RANSAC has been proven to be remarkable by experimental evaluation. In this paper, RANSAC algorithm was used to eliminate the false matches. In the fusion stage, fade-in and fade-out method is applied to smooth seams which exist in the stitched image, but the effect is not obvious, so the adaptive Gamma correction method is used to weaken illumination effect on image quality. Finally, a stitched image is obtained, which contains more information than each of the original images.

7.2.2.3 Image Fusion

According to the transformation matrix between images, the corresponding images can be transformed to determine the overlapping areas between images, and the fused images can be mapped to a new blank image to form a mosaic image. However, due to the difference of brightness between the input images caused by the exposure parameters automatically selected by ordinary cameras when taking photos, obvious changes of light and shade appear at both ends of the stitching line. Therefore, image fusion of stitched images is still needed. Image fusion is to eliminate the discontinuity of the intensity or color of the image, and smooth the intensity and color of the image at the stitching point. In order to obtain high quality mosaic images, image fusion algorithm should satisfy two conditions: smooth transition between images, no seams, and no loss of original image information. Commonly, the image fusion methods are the average method, the gradual-in–out method and the median filtering fusion method. In this approach, a fast and simple gradual-in and gradual-out algorithm is used to deal with the splicing seam problem.

Using the above stages to realize image mosaic, the stitched image has obvious seams in overlap region, for we ignored the difference of illumination. The fade-in and fade-out method is used in this paper. This method is based on the weighting method, and its weight is determined by the distance from the pixels to the border of the overlap region.

$$I(x, y) = d_1 I_1(x, y) + d_2 I_2(x, y) \tag{7.4}$$

where d_1 is the distance between the pixel to the left margin, d_2 is the distance of the pixel to the right margin. I, $I1$ and $I2$ is the pixel of point in the image. From right to left in the overlap region, the percent of $I1$ grades from 0 to 1, and $I2$ grades from 1 to 0, as to transition smoothly in the overlap region, but the effect is not obvious. So the adaptive Gamma correction method is proposed to weaken illumination effect on image quality. A complete stitched image can be attained by the above steps.

7.3 Experimental Results and Discussions

Morphological characteristic is one of the important factors of grass seed variety recognition. However, in grass seed recognition experiments, it is difficult to acquire high resolution images of both details and overall appearance of the grass seed due to the limitations of microscope and the unsmooth specimen surface at the same time. The purpose of writing this article is to solve this problem.

The images used in this paper are grass seed data recorded by the microscope and CCD camera. 12 images are established as Image sequence, and the images are coded with the Fig. 7.5b, such as Image1, Image 2, etc. Here are some examples with uniform size of 512×384. The image (a) is an overall image captured by a common digital camera with a macro lens whose resolution of the local details is often insufficient. The local microscope image sequence (b) show detailed different sections of a seed. It is clear that there is an overlap field between the two images, which is crucial to feature matching based on SIFT, and can be used as images similarity to exact feature points. If we integrate information from the image sequence into a resultant image, it will improve the efficiency and facilitate the work for experts who want to identify the grass seed.

Image should be pre-processed before the feature matching, this is important for different pictures with different focus and brightness. After image fusion stage, experiments show that the seams in the image still exist. In response to this problem, an adaptive Gamma correction method is applied to weaken illumination effect on image quality in this paper. Firstly, a mapping between pixel values and Gamma values is built. The Gamma values are then revised using two non-linear functions to prevent image distortion. Finally, pixels are corrected adaptively using the readjusted

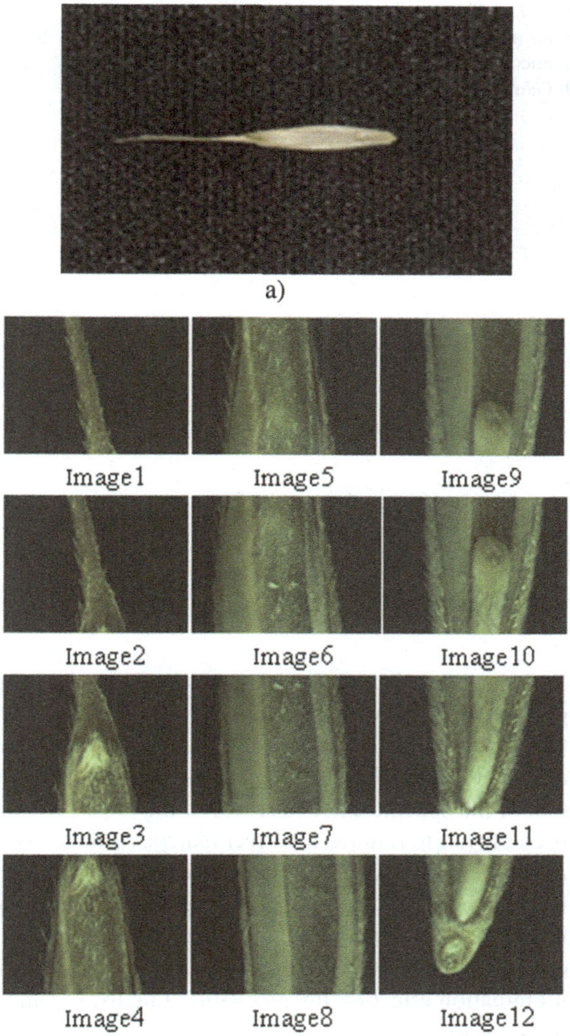

Fig. 7.5 Forage seed image (Elymus nutans Griseh), **a** overall image of the seed, **b** micro-image array to be mosaiced

Gamma values. The panorama grass seed image after stitched is shown in Fig. 7.6. The obvious seams that are smoothed by Gamma correction method from Fig. 7.6a, b can be found.

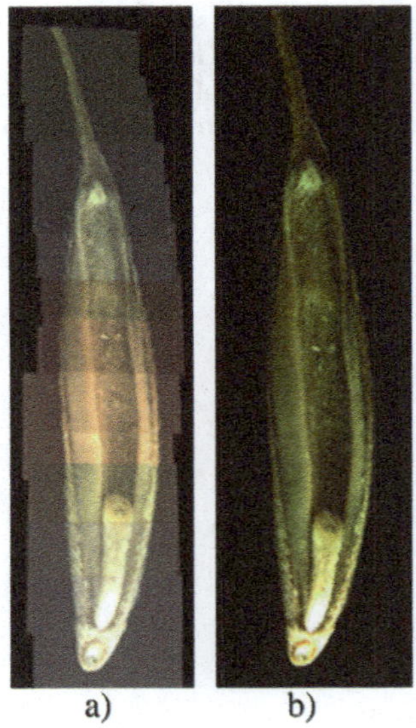

Fig. 7.6 Panorama image after image mosaic, **a** uncorrected by gamma, **b** Gamma corrected

7.4 Summary

In this chapter, we investigate image mosaic algorithms and its application in integration of gramineous grass seed micro-images. The SIFT algorithm is employed to extract feature points, define descriptors and obtain initial matches, but some incorrect matches greatly affect the registration accuracy. Therefore, in consideration of the specific characteristics of the grass seed image, the RANSAC algorithm is employed to remove incorrect matches effectively. The refined registration result is obtained through adjusting the registered image on the reference image in the neighborhood of the overlapping region with tiny steps. Image fusion based on the fade-in and fade-out method is used to smooth seams of the stitched image, but the effect is not obvious, so the adaptive Gamma correction method is used to weaken illumination effect on image quality. Finally, a complete stitched image containing more information can be obtained.

References

Feng, Z. J., M. Luo, S. Chang, Y. Li, and K. Jun. 2009. Automatic Cartridge Case Image Mosaic Using SIFT and Graph Transformation Matching[C]. Proceedings of 2009 IEEE International Conference on Intelligent Computing and Intelligent Systems (ICIS2009). 04: 371–375.

Fischler, M. A., and R. C. Bolles. 1981. Random sample consensus: a paradigm for model fitting with applications to image analysis and automated cartography[C]. Communications of the ACM. 24(6):381–395.

Fischler, M. A., and Bolles, R. C. 1987. Random sample consensus: a paradigm for model fitting with applications to image analysis and automated cartography—sciencedirect. Readings in Computer Vision, 726–740.

Hou, J. H. 2007. Research on image denoising approach based on wavelet and its statistical[C]. Huazhong University of Science and Technology. (in Chinese).

Lowe, D. G. 2004. Distinctive image features from scale-invariant key points[J]. International Journal of Computer Vision. 60(2):91–110.

Ma, L., J. H. Wang, K. Q. Wang, and D. P. Zhang. 2010. Adaptive gamma correction method of Iris image based on characteristic pattern[J]. Journal of Yanshan University. 02:173–179. (in Chinese).

Mikolajczyk, K. , and Schmid, C. 2005. A performance evaluation of local descriptors. IEEE Transactions on Pattern Analysis and Machine Intelligence, 27(10):1615–1630.

Ning, L. N., and X. Pan. 2013. Image mosaic algorithm and its application to the microimage of grass seeds[C], 2013 International Conference on Information Science and Computer Applications (ISCA2013), 179–184.

Song, Z. L. 2010. Research on image registration algorithm and its applications[D]. Fudan University.

Wang, J., J. Shi, and X. X. Wu. 2008. Survey of image mosaics techniques[J]. Application Research of Computers. 07:1940–1943 + 1947.

Wendy, A., F. Yann, E. Francisco, M. P. M. Elena, E. R. Arturo, and A. L. Miguel. 2009. A robust graph transformation matching for non-rigid registration[J]. Image and Vision Computing. 27:897–910.

Chapter 8
Digital Information Platform of Grassland and Forage Based on Computer Vision

8.1 Digital Information Platform of Grassland and Forage

The research of grassland digitalization is to manage grassland information in a digital mode. The purpose of designing and implementation of a digital platform of grassland and forage is to provide a public communication path for scientific researchers, grassland workers, managers, students and herdsmen, so as to realize real-time update of data, resource sharing and grassland management. It has also been implemented a convenient and efficient management mode for grassland management. At present, similar websites in China provide some information and related knowledge about grassland, such as the first Chinese grassland resource information system www.grassland.net.cn, which is based on 3S technology. In 2001, it has been published by the National Key Basic Research and Development Planning (973) Project Department, the Key Open Laboratory of Grassland Resources Ecology of the Ministry of Agriculture and the Grassland Research Institute of the Chinese Academy of Agricultural Sciences. It provides a platform for scientific researchers to acquire and publish basic data, in which 10 ecological information of forage distribution area can be quickly queried. The query method is fast and simple, which lays a solid foundation for the construction of digital grassland. Because the information system has designed and developed with Java, it is limited by the development environment when accessing it. It is necessary to install the JRE module (Java Runtime Environment) of the Java running environment. Although the Web provides links to download components, these additional steps will cause a great deal of time consumption of the users. Thus it is often unacceptable to users, which limits the scope of the system. The system provides a large amount of remote sensing data as well. The disadvantage of the system is that it browses slowly. In addition, the system does not have images of grassland and forage acquired by digital camera or it can't process and display images. Moreover, it can't effectively collect daily data obtained by researchers. It lacks the interactive function with users. Therefore, this research designs and implements a grassland digital platform based on computer vision. At the same time, it integrates image processing and pattern recognition functions into

X. Pan et al., *Computer Vision based Identification and Mosaic of Gramineous Grass Seeds*, https://doi.org/10.1007/978-981-16-3501-4_8

the platform. The advantage of the system is to realize real-time image processing and recognition, improve the utilization efficiency of grassland images and the level of digital grassland management.

8.1.1 Main Development Tools and Models for Platform

The image processing function module of the platform is mainly realized by MATLAB, such as image mosaic. The platform uses JSP to edit dynamic scripts, applies open source framework Struts, Hibernate, and Spring to develop (Chen 2006), and employs Mysql to realize seamless connection between JDBC driver and database.

In Web browser,the function of image processing and recognition is mainly based on the hybrid programming technology of Java and matlab.Firstly, it is packaged into Java readable Jar file by Matlab Builder JA in Matlab, and then parameters are transferred and invoked in Java language. Secondly, parameter transfer and function call are implemented in Java. In order to facilitate clients to use those functions, GUI interface of image mosaic algorithm is realized by GUIDE in Matlab. The GUI files are converted into.exe file, which is easy to download and run. This file can run independently without Matlab environment.

Development tools and models of the platform are summarized as follows.

Matrix is the abbreviation of Matrix Laboratory. It has been launched by MathWorks since 1984. Now, it has become an internationally recognized excellent engineering application environment through continuous improvement and development. In the field of image processing, MATLAB has developed powerful modules and toolboxes, which contain a large number of image processing functions. When realizing the function of image processing and recognition, the amount of code is much less than that of Java. The probability of error is small and it is easier to maintain. In order to make the function of MATLAB run independently of the software environment, it provides two methods: one is that users can apply Lcc compiler to convert MATLAB function or graphical user interface into EXE file. The other is to package functions into one or more jar packages by the function of MATLAB Builder JA. Through mixed programming of Java and MATLAB, MATLAB functions can have more application platforms. In addition, MATLAB software provides a way of interaction mode between users and computer programs, called GUI (Graphical User Interface). Users can directly run packaged.exe files without understanding internal operations (Deng 2012).

JSP is a dynamic technology standard, which is initiated by Sun Microsystems and participated by many companies. In traditional HTML files (*. htm, *. html), JSP pages are formed by adding Java program fragments (Scriptlets) and JSP tags. JSP page execution process is: when visiting a JSP page for the first time, the file is initially translated into a Java source file by JSP container, which is actually a Servlet. Then, it is compiled and generated the corresponding bytecode file. Class. Next, like other Servlets, it is handled by Servlet containers. Servlet container loads this class, processes requests from customers and returns results to customers.

Java language is a pure object-oriented programming language introduced by Sun Microsystems in May 1995, which is suitable for network programming. It integrates the advantages of C++ and other languages. For example, Java does not support operator overloading, multi-level inheritance and extensive automatic mandatory features. The function of automatic garbage collection in memory space is added. However, Java language lacks in image processing and data analysis. Due to multiple loops, matrix operation will run slowly. Therefore, the hybrid programming technology of Java and Matlab can combine the advantages of MATLAB in image processing and Java has multi-threading mechanism to facilitate real-time interaction on the network. Finally, image processing based on browser is realized.

Java is a pure object-oriented programming language introduced by Sun Microsystems in May 1995, which is suitable for network programming. It integrates the advantages of C++ and other languages. For example, Java does not support operator overloading, multi-level inheritance and extensive automatic mandatory features. The function of automatic garbage collection in memory space is added. However, Java language lacks in image processing and data analysis. Due to multiple loops, matrix operation will run slowly. Therefore, the hybrid programming technology of Java and Matlab can combine the advantages of MATLAB in image processing and Java has multi-threading mechanism to facilitate real-time interaction on the network. Finally, image processing and recognition based on browser are realized.

J2EE (Java 2 Platform Enterprise Edition) is a set of technical architecture that simplifies the development of enterprise solutions, deploys and manages related complex problems by Java 2 platform. It is suitable for creating server applications. Its core is a set of technical specifications and guidelines, including many components, service architectures and technical levels. They have common standards and specifications, which can simplify and standardize the development and deployment of application systems. This makes the different platforms based on J2EE architecture have good compatibility, security and reuse value. On this basis, the pasture image mosaic system has high scalability, portability and platform independence. Users can complete high quality image mosaic operation without installing MATLAB on the client, which not only improves the efficiency of programs, but also saves the cost of multiple deployments.

MVC (Model View Controller)is a framework model of software design, which mainly includes three parts: Model, View and Controller. These three parts can not only deal with their own tasks, but also cooperate to complete the work submitted by users. SSH framework is the integration of multiple frameworks (struts + Spring + hibernate). It is a popular framework for open source integration of Web applications, which is mainly used to build flexible and scalable multi-level programs.

Struts2 takes Servlet and JSP tags as part of its implementation. This framework inherits the characteristics of MVC. According to the characteristics of J2EE, it changes and extends. The coupling between business logic interface and data interface is weakened, which makes the view layer more changeable. In addition, Struts 2 framework also has page navigation function, which can connect all parts of the whole system through a file configuration. It facilitates system maintenance. Spring is an open source framework created by Rod Johnson to solve the complexity of enterprise applications. Its main task is to refine the business layer, that is, to reduce the degree of coupling at a deeper level. It can help programmers to complete dependency concerns between components, so that dependencies can be minimized. Hibernate is an open source object relational mapping framework based on Java, which encapsulates JDBC very lightweight objects. Hibernate can replace CMP in the application of EJB (Enterprise JavaBeans) J2EE architecture, and can be integrated into J2EE system as a persistence layer framework. Hibernate can replace CMP in J2EE architecture of EJB application, and be integrated into J2EE system as a persistence layer framework.

SSH is a new framework that integrates Struts, Spring and Hibernate.The integrated SSH framework system can be divided into four layers: presentation layer, business logic layer, data persistence layer and domain module layer (entity layer).Such layering can help system developers build Web applications with clear structure, good reusability and easy maintenance in a short time. The three sub-frameworks include: Struts as the overall infrastructure of the system, mainly responsible for the separation of MVC; Hibernate framework to support the persistence layer; Spring is responsible for the management of struts and hibernate. The functions of the three sub-frameworks in the new SSH frameworks are as follows: as the overall infrastructure of the system, Struts mainly responsible for the separation of MVC; Hibernate framework support the persistence layer; Spring is responsible for the management of struts and hibernate.

8.1.2 Main Function Modules of the Platform

The purpose of designing and realizing grassland digitalization platform is to provide a public communication place for scientific researchers, grassland workers, managers, students and herdsmen, and to realize real-time data updating, resource sharing and grassland management. It has also been implemented a convenient and efficient management mode for grassland management and improved the digital management level of grassland. According to the requirement analysis, users are divided into three categories: visitors, members and administrators with different rights. The overall framework of the platform is shown in Fig. 8.1.

In general, the platform can be divided into two modules: the front and the background. The front desk includes information management, membership management, and image processing and recognition sub-modules. Background includes member management, administrator management, image management, software management; system management and data update sub-modules.

8.2 Key Technological Problems and Solutions of the Platform

8.2.1 Hybrid Programming of Java and Matlab

Beginning with, Matlab Builder JA (matlab builder for Java) function has been inserted since MATLAB Version 6.0. With this function, the programmer can convert

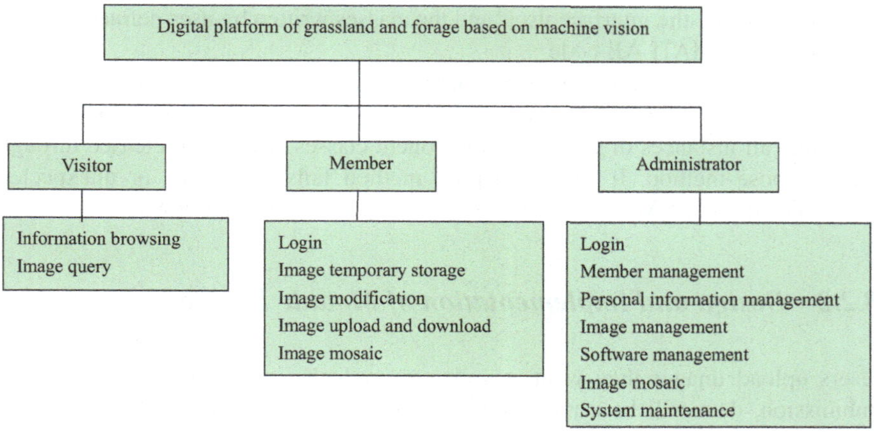

Fig. 8.1 Block diagram of digital information platform for grassland and forage

matlab programs into corresponding number of Java classes. These Java classes can be packaged into a jar package, a Java component, while MATLAB functions are compiled into methods in a Java class. These Java classes can be integrated into Java programs without installing MATLAB and deployed to desktop computers or network servers, which can be accessed by multiple users through Web browsers (Li 2009, Ren et al. 2009, Wang and Yang 2012, Zhou 2013).

The workflow of the MATLAB Builder JA tool is briefly introduced as follows.

According to the javabuilder. jar file,Matlab Builder JA converts Matlab programs into a Java class and packages it into a jar package that can be called by Java. Note that the javabuilder. jar package runs on the condition that the MCR matching the version is installed on the computer at the same time. It should be noted that javabuilder. jar runs on the condition that the MCR matching the version is installed on this computer.

When a class contained in javabuilder. jar is instantiated for the first time, a series of events occur:

Load javabuilder. jar, which is dependent on many classes.

(1) Static initialization of related classes triggers the loading of a series of shared libraries (including the compiler runtime MCR of MATLAB).
These shared libraries will implement some local methods, which are the bridge between MATLAB Builder JA components and MCR implementation.
After the shared library is loaded, MATLAB runtime will be initialized by C++ class of mcrInstance.
(2) The construction of MCR Instance will lead to the initialization of many environment subsystems, including MATLAB language runtime. Such a subsystem is Java interface in MATLAB, which allows MATLAB programs to call Java code directly.
(3) Attach current thread uses Java interface of MATLAB to create a class loader to load all classes required by MATLAB code. These include the basic classes

required by the interface itself and the classes written by user-defined directly imported MATLAB code.

(4) The static constructor in javabuilder installs a closing hook to terminate MCR thread and release resources. This process enters closed state through JVM and all instances of generated component classes have been released through dispose method. If calling dispose method fails, undefined or unexpected behavior of JVM occurs when original threads run to shut down.

8.2.2 Design and Implementation of Module Interface

Users upload images they want to splice through the interactive interface. After submission, data will be transmitted to Java component built by MATLAB Java Builder. Java component runs the image stitching program by calling Matlab compiler runtime (MCR) and returns results. Then, the server-side program passes results to user's browser.

The preprocessing and mosaic of forage microscopic images are realized by matlab. The program is named forage_mosaic as a Matlab function. Input parameters of the function are names of two images, which are want to be stitched. The output is the image processed by stitching function. The algorithm can be effectively separated from the system by MATLAB language. When algorithms need to be improved, it only needs to modify MATLAB code and repackage, without changing other parts of the system. In this way, algorithm developers do not need to pay attention to design the system, but to design algorithms. Similarly, system developers do not need to be concerned with designing algorithms, but to design systems.

Because the algorithm modules are written in Matlab, if they are required to deploy in the system, the tool of MATLAB builder JA need to be used to package Matlab programs of algorithms into a Jar library file. The specific process is as follows:

(1) Start MATLAB software, input deployment on the command line and call out Deployment Tool.
(2) Set up the location and name of the project in deployment window interface of Matlab Builder JA (named mosaic in this article).Type should select Java Package and click OK button. The system generates a Java package called mosaic.prj.
(3) Click Add class in Java Package window to add Java class and change class name Class1 to mosaic. Matlab programs of the algorithm are added to the mosaic folder, and the matlab program file is named forage_mosaic.
(4) The package can be packaged by package button on the toolbar.In this way, a mosaic.jar file can be generated, which can be used by other developers on different machines and can be invoked in Web programs.

After encapsulatingMATLAB programs, matlab programs can be called in Java. Data transfer between MATLAB program and Java program is inevitable in invocation. But due to the different data type structure between them, the focus of work shifts to how to transform and transfer data between them.

In order to use the Java classes generated by MATLAB Builder JA, it is necessary to import MATLAB variable library and component classes into Java code through import functions. For example:

```
import com.mathworks.toolbox.javabuilder.*;
import componentname.classname; or import componentname.*;
```

In order to enable data conversion and transmission between Java and matlab, Java Builder provides a package of com.mathworks.toolbox.Javabuilder.MWArray, which contains a series of classes inherited from MWArray, each representing a MATLAB data type. Applying classes in this package, programmers can exchange data between Java applications and encapsulated Matlab functions. The details are as follows:

(1) The Matlab function transfers data to Java. For example, when a string of type Matlab char need to passed, Matlab program returns an MWCharArray object to Java. Returned data is not converted to Java data type. If Java data types need to used, the corresponding to Array method that inherits from the MWArray class must be used.
(2) Java program transfers data to MATLAB function. For example, when Java string need to passed to a Matlab function, just need to convert Java string to an MWCharArray object.

8.2.3 Design and Implementation of GUI for Image Mosaic

In this platform, the image mosaic function is realized by Matlab. The Matlab codes are transformed into Java classes by Matlab Java Builder toolkit. In MyEclipse, the programs are written to call the Servlet and deploy it to the Tomcat server. By responding to image processing requests of client browser, the processing results of MATLAB GUI programs are returned to client through HTTP protocol by MVC design mode, so as to realize image mosaic taking advantage of computer network.

Graph user interface (GUI) file of image mosaic is an object-oriented visual interface designed by graphical user development environment GUIDE contained in MATLAB tools. The fulfillments of background call back for all the listed functions in the interface, so as to complete its design (Fig. 8.4). The GUI file can accomplish the following functions: (1) selecting two local images to be spliced; (2) preprocessing the selected images; (3) image mosaic for the pre-processed images; (4) saving the spliced image into the local path; (5) selecting the region of interest of the mosaic image to remove the redundant area in the image.

The disadvantage of the graphical user interface created by Matlab software is that it can only run on a computer equipped with Matlab software. To solve the problem, Matlab Builder JA tool was used to convert the GUI platform of image mosaic written in Matlab code into an executable file. A smaller component of Matlab is required to install on the computer to execute the program.

8.3 Implementation of the Platform

The user interface of the platform is shown in Fig. 8.2. Visitors can register to be members of the platform via accessing the home page. Information can be browsed and queried from the platform as well. In addition, members have the privileges of temporary storage, modification, upload, download, image processing and identification, in which the parameters can be set to meet personalized needs. At the same time, members can also download the exe file of image processing, which can be used for image mosaic and processing on the client side even when the network service are sometimes unavailable in the grassland. The administrator access the backstage via management page, owning the right to choose, delete the pictures the user uploaded, manage the users, system maintenance and data update authority.

Members are entitled to access the image processing and recognition module, and the image Mosaic page is shown as Fig. 8.3. In the image Mosaic webpage,

Fig. 8.2 Homepage of the platform

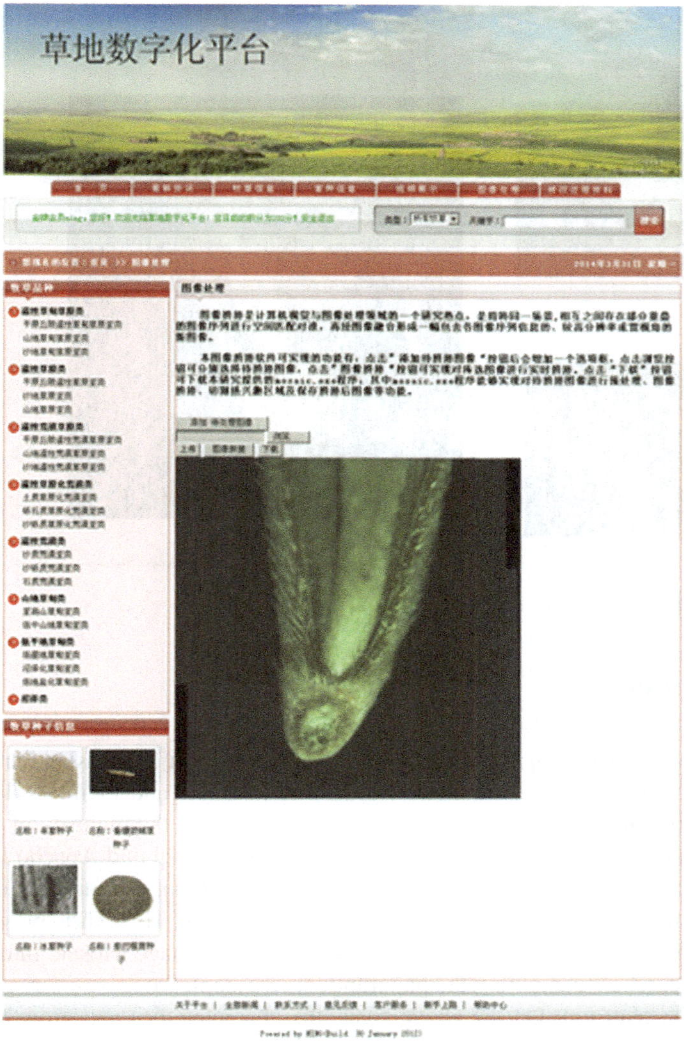

Fig. 8.3 Webpage for image mosaic

the member user can realize the following functions: upload and download forage images; upload the images to be spliced, image mosaic online and saving spliced image. Download the executable program of image mosaic mosaic.exe, which can realize the pre-processing, image Mosaic, ROI segmentation and file saving after image mosaic from the local client without network (Fig. 8.4).

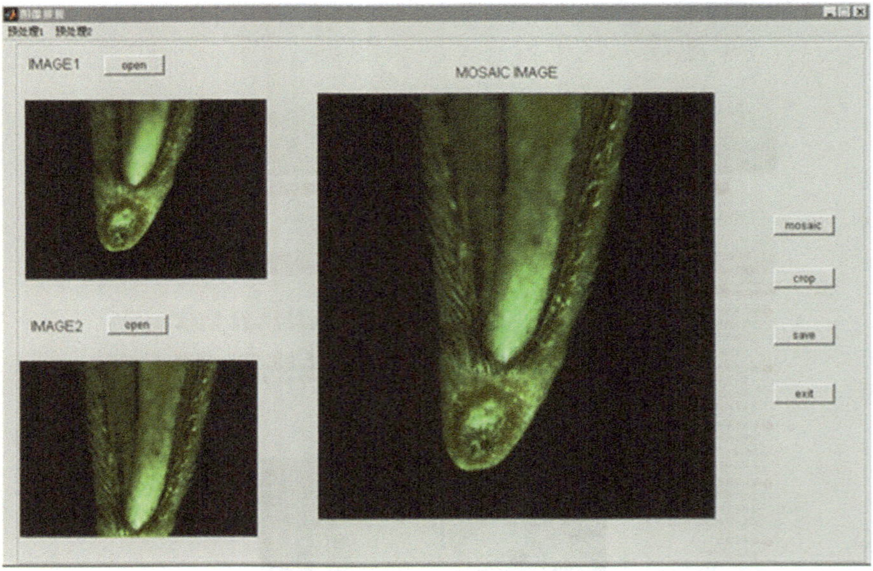

Fig. 8.4 GUI of image mosaic

8.4 Summary

In this chapter, we design and realize the digital platform of grassland and forage, which provides basic information of grassland herbage, and facilities the professional information exchange and knowledge learning for the majority of grassland researchers, officers in grassroots and students. JSP, Mysql and Apache have been used to achieve the basic functions. Different from ordinary digital platforms, this system emphasizes the function based on computer vision. The hybrid programming of Java and Matlab integrates the realized function of image Mosaic into the platform. Researchers can obtain the panoramic microscopic images of forage or seeds through the platform. On the above basis, the characteristics and growth environment of forage can be further acquired, and a relatively intact understanding can be obtained to facilitate the management of grassland and forage.

References

Chen, Z. J. 2006. The research and implementation of web application system based on Struts, Spring and Hibernate[D]. Liaoning Technical University. (in Chinese).

Deng, C. R. 2012. Based on the MATLAB GUI multi-functional calculation system design and realization[D]. Nanchang University. (in Chinese).

Li, W. Q. 2009. Application of mixed programming based on Java and Matlab in image process[J]. China Computer & Communication. 11(10):108–111. (in Chinese).

Ren, W. J., W. Wang, S. H. Ma, and H. M. Chen. 2009. Research and implementation of Mixed-Language programming based on MATLAB and JAVA[J]. Measurement & Control Technology. 28(1):77–79 + 82. (in Chinese).

Wang, W., and L. P. Yang. 2012. Mixed programming of Java and Matlab and its application[J]. Journal of Changchun University. 22(10):1186–1189. (in Chinese).

Zhou, X. J. 2013. Research of Java-Matlab mixed programming methodology[J]. Electronic Design Engineering. 21(2):16–18 + 23. (in Chinese).

Appendix
Gramineous Grass

A. Elymus L.

A1. Elymus Dahuricus Turcz

[Features]

Stalk sparse scattered, erect, 70–140 cm tall, geniculate at the base. Sheath glabrous; leaf-blades flat, scarce and involute, coarse at the top and glabrous beneath, pink green, 15–25 cm long, and 5–9 (12) mm wide. Spike erect, tight, 14–18 cm long, 5–10 mm wide; the margin of spike-stalk has ciliolum, each internode of middle part has 2 spikelets, and 1 spikelet near the apex and at the base; spikelet green, and turns into olive drab when mature, 10–15 mm long, with 3–5 florets; glumes lanceolate or linear-lanceolate, 8–10 mm long, 5 mm-long awn at the apex, 3–5 coarse veins; lemmas lanceolate, with 5 conspicuous veins at the top, dense and short pilose, the first lemma 9 mm long, awn coarse, 10–20 mm long, unfolding outward when mature; palea as long as lemma, flat at the apex, ciliiform on the ridge, and scarce short hairs between ridges.

[Habitat]

Most of them are found on hillside, grasslands and roadsides.

[Use]

Forage use (Figs. A1.1, A1.2, A1.3, A1.4, A1.5 and A1.6).

X. Pan et al., *Computer Vision based Identification and Mosaic of Gramineous Grass Seeds*, https://doi.org/10.1007/978-981-16-3501-4

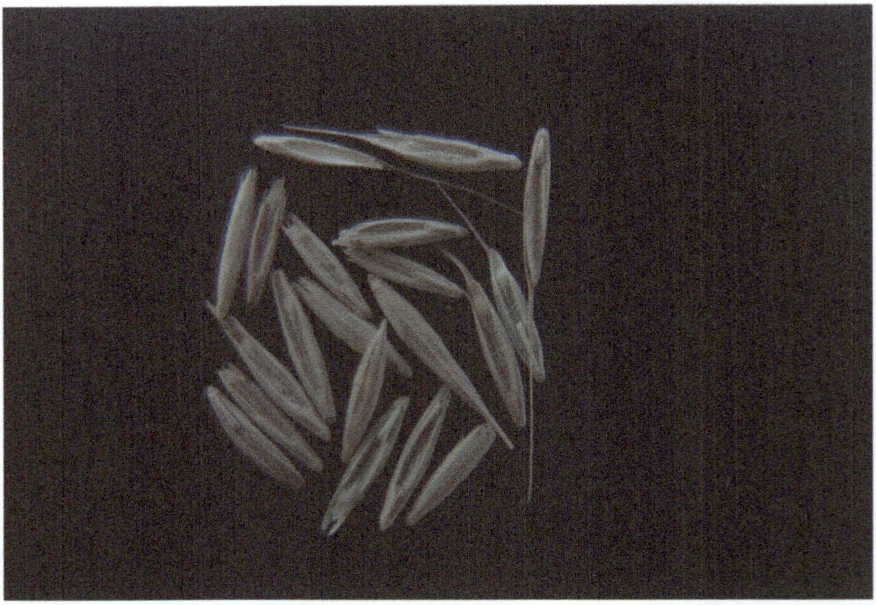

Fig. A1.1 Seeds (*Elymus dahuricus* Turcz.)

Fig. A1.2 Field and samples (*Elymus dahuricus* Turcz.)

Fig. A1.3 Plant (Elymus dahuricus Turcz.)

Fig. A1.4 Leaves (Elymus dahuricus Turcz.)

Fig. A1.5 Spike 1 (Elymus dahuricus Turcz.)

Fig. A1.6 Spike 2 (Elymus dahuricus Turcz.)

A2. Elymus Nutans Griseb

[Features]

Perennial clump herb. Scales 50–70 cm tall, upright, slightly knee-curved at base part. Leaf sheath pilose at base and roots. Leaves flat, sparsely pilose on the top, rough or smooth on the bottom, 6–8 cm long and 3–5 mm wide. The spikes are relatively compact, usually tortuous and drooping at the tip, 5–12 cm long, with rough cob edges or small cilia. Nodes 1 and 2 have no developed spikelet at the base, green fish with spikelet, purple after maturity, usually 2 at each node and only 1 near the top and lower nodes, more or less on the cob 1 Side, nearly sessile or very short stalk, 12–15 mm long, with 3–4 florets. Glumes 4–5 mm long, oblong, 2 glumes equal, apex acuminate or short awn 1–4 mm long, 3–4 veins, obvious and rough veins; lemma is long lanceolate, 5 veins, veins inconspicuous at base, all covered with tiny short hairs. First lemma 10 mm long, top extended into an awn. The awn rough, curved outwards or slightly spread out, 12–20 mm long. Inner lemma and external lemma equal length, apex blunt or truncated, cilia on the ridge, hair gradually disappears toward the base, ridge covered with tiny short hairs.

[Habitat]

Grasslands, hillside roads, forest margins.

[Use]

Forage use (Figs. A2.1 and A2.2).

Fig. A2.1 Seeds (Elymus nutans Griseb)

Fig. A2.2 Scattered seeds (Elymus nutans Griseb)

A3. Elymus Cylindricus (Frmch) Hinda

[Features]

Perennial clump herb. Stems thin, 40–80 cm high. Sheath glabrous, leaves flat, curled inward after drying, 5–12 cm long, 5 mm wide, rough on the top and smooth on the bottom. Spikes erect, narrow and thin, 7–14 cm long, 5 mm thick, except for each node near the apex with only 1 spikelet, the other nodes have 2 spikelets; the edges of the cob are small cilia; spikelets green or banded purple, 9–11 mm long (except awns), 2–3 florets, 1–2 florets develop; glume-lanceolate to linear-lanceolate, 7–8 mm long, 3–5 veins, with obvious veins rough, apex acuminate or with a short awn up to 4 mm; the outer float lanceolate, all covered with tiny short hairs, first outer float 7–8 mm long, 5 veins, apex rough, upright or slightly spread, 6–13 mm long; the inner scale and the outer float are the same length, tip blunt and round, ridges cilia, and the ridges are covered with tiny short hairs.

[Habitat]

Hillside, roadside.

[Use]

Forage use (Figs. A3.1 and A3.2).

Fig. A3.1 Seeds (Elymus cylindricus (Frmch) Hinda)

Fig. A3.2 Scattered seeds (Elymus cylindricus (Frmch) Hinda)

A4. Elymus Sibiricus Linn

[Features]

Perennial sparse grass. The fibrous roots dense and well-developed, penetrate deep into the soil, strong tillering power. Stem 90–98 cm tall, upright or slightly inclined at the base. Leaves flat, involute, 11-17 cm long, 5-7 mm wide, with abundant leaf volume and large leaves. Spikes loosely curved and drooping, 18–21 cm long, 15–20 mm wide. The lemma lanceolate, densely covered with microhairs, awns are slightly expanded or recurved, 15 mm long, inner lemma and external lemma equal length. Caryopsis oblong and easily fall off. The yield of hay and seeds is higher. Salt and drought tolerance. Resistance to cold, pests and diseases. Wide adaptability.

[Habitat]

Roadsides and hillside.

[Use]

Forage use (Figs. A4.1 and A4.2).

Fig. A4.1 Seeds (Elymus sibiricus Linn.)

Fig. A4.2 Field and samples (Elymus sibiricus Linn.)

A5. Roegneria Kamoji Ohwi

[Features]

Perennial herb, pole upright stem or inclined base, 30–100 cm tall. The outer edge of the leaf sheath often ciliated; leaf blade flat, 5–40 cm long, 3–13 mm wide. Spikes 7–20 cm long, curved or drooping; spikelets green or purple, 13–25 mm long (except awns), containing 3–10 small flowers, glumes ovate-lanceolate to oblong-lanceolate, apex sharp to short awn (2–7 mm long), broad membranous margin, first glume length 4–6 mm, second glume length 5–9 mm, lemma lanceolate, with a wide membrane. The qualitative edge, the back and the base plate nearly glabrous or only the two sides of the base plate have very small short hairs, upper part has obvious 5 veins, the veins slightly rough, first lemma 8–11 mm long, apex extends into a awn. Awns rough, straight or slightly tortuous on the upper part, 20–40 mm long; palea is about the same length as the lemma, apex blunt, ridge prominently winged, flange has fine cilia.

[Habitat]

Most of them are found on hillsides and moist grasslands at an altitude of 100–2300 m.

[Use]

Forage use, medicinal use (Figs. A5.1, A5.2, A5.3 and A5.4).

Fig. A5.1 Seeds (ARoegneria kamoji Ohwi)

Fig. A5.2 Scattered seeds (ARoegneria kamoji Ohwi)

Fig. A5.3 Field 1 (Roegneria kamoji Ohwi)

Fig. A5.4 Field 2 (Roegneria kamoji Ohwi)

A6. Roegneria Tunczaninovii(Drob.)Nevski Var.Macrathera Ohwi

[Features]

The plant has roots; stalk is thicker, sparsely clumped, 60–80 cm high, base stem about 3 mm. The upper leaf sheath smooth and glabrous, lower part often has downed hairs; leaves soft and flat, 9–20 cm long (tiller leaves up to 26 cm), 3–8 mm wide, covered with short hairs on the top and glabrous on the bottom. Spikes erect, 9–15 cm long, 7–13 spikelets, often on one side; spikelets 18–25 mm long (except for the awn), yellow-green or slightly blue-purple, 5–7 small flowers; glums lanceolate, apex or acuminate, 3–5 thick veins, 1–2 shorter, thin veins, rough on the veins, two glumes are equal in length, about 10–15 mm long, equal to or the second glume is longer than the first lemma; lemma lanceolate, tiny bristles all over the body, obvious 5 veins on the upper part, first lemma 10–12 mm long, apex 2.7–4.3 cm long; lemma the upper part of the ridge has short hard cilia, tip blunt or slightly concave. The upper part of the back has short hard cilia.

[Habitat]

Roadsides.

[Use]

Forage use (Fig. A6.1).

Fig. A6.1 Field (Roegneria tunczaninovii(Drob.)Nevski var.macrathera Ohwi) https://baike.baidu.com/pic/%E5%A4%A7%E8%8A%92%E9%B9%85%E8%A7%82%E8%8D%89/10915133/1/34fae6cd7b899e5160877b2540a7d933c8950df1?fr=lemma&ct=single

A7. Roegneria Varia Keng

[Features]

The scales sparse, erect or slightly kinked at the base, 17–40 (60) cm high. The leaf sheath smooth; leaves hard and usually curled, glabrous, or sparsely pubescent above (rarely on both sides of the basal leaves), 3.5–15 cm long and 1–3 mm wide. Spikes erect, 3.5–10 cm long (except awns); spikelets arranged on both sides, green or purple at maturity, containing 3–5 small flowers; glumes are oblong-lanceolate, slightly asymmetrical, apex sharp. With a small pointed head, with 3–5 veins, rough on the veins, first glume 4.5–8 mm long, second glume 6–10 mm long; outer floating lanceolate shape, back glabrous or the edges, the upper part attached with fine hairs, The upper part has obvious 5 veins, first outer float 7–8.5 mm long, apex 7–17 mm long; inner release is the same length as the barnyard, apex slightly concave, spine is above the middle and has cilia, spine also appended with micro hairs .

[Habitat]

Hillside.

[Use]

Forage use (Fig. A7.1).

Fig. A7.1 Plant (Roegneria varia Keng) https://baike.baidu.com/pic/%E5%A4%9A%E5%8F%98%E9%B9%85%E8%A7%82%E8%8D%89/7085937/1/d788d43f8794a4c2198fb8190df41bd5ad6e3957?fr=lemma&ct=single

B. Agropyron Gaertn

B1. Agropyron Cristatum (L.) Gaertn

[Features]

Perennial herb, dense fibrous root. Stalk sparsely fasciculate, short-pilose at the top, 30–60 (75)cm tall, 2–3 internodes, sheath stem-clasping, shorter than internodes; leaf-blades 5–20 cm long, 2–5 mm wide, hard and coarse, usually involute. Spike stout, 2.5–5.5 cm long, 8–15 mm wide; spikelet parallelly arranged in two lines, regular, pectinatory, containing 4–7 flowers, 10–13 mm long, about 3 mm wide; glume elliptic-oblong; the first glume 2–3 mm long, the second glume 3–4 mm long, long awn; lemma 6–7 mm long, thin and narrow edge with bristles, awn 2–4 mm long; palea as long as lemma, abruptly cuspidate at the apex, primary veins on the reverse side covered with short bristles; anthers yellow, about 3 mm long. Caryopsis 4 mm long.

[Habitat]

It usually grows on dry grasslands, hillsides, hills and sandy land.

[Use]

Forage use, medicinal use (Figs. B1.1, B1.2, B1.3, B1.4, B1.5 and B1.6).

Fig. B1.1 Seeds (*Agropyron cristatum* (L.) Gaertn.)

Fig. B1.2 Field and sample (Agropyron cristatum (L.) Gaertn.)

Fig. B1.3 Field (Agropyron cristatum (L.) Gaertn.)

Fig. B1.4 Plant (Agropyron cristatum (L.) Gaertn.)

Fig. B1.5 1-Spikelet (Agropyron cristatum (L.) Gaertn.)

Fig. B1.6 2-Spikelet (Agropyron cristatum (L.) Gaertn.)

B2. Agropyron Cristatum (L.) Gaertn. Var. Pectiniforme (Roem. et Schult.) H. L. Yang

[Features]

It grows longevity sparse clump grass for many years. The fibrous roots are well-developed, densely growing, with a sand jacket. The stalk is erect, the plant 75–81 cm tall. Leaves long, hard and rough, often involute. The spikes are erect, 6–7.5 cm long, 10–18 mm wide. The spikelets are arranged horizontally and are dentate. The glumes and lemmas are all smooth and glabrous or sparsely covered with short bristles of 0.1–0.2 mm.

[Habitat]

Hillside.

[Use]

Forage use (Figs. B2.1, B2.2 and B2.3).

Fig. B2.1 Scattered seeds (Agropyron cristatum (L.) Gaertn. var. pectiniforme (Roem. et Schult.) H. L. Yang)

Fig. B2.2 Field and samples (Agropyron cristatum (L.) Gaertn. var. pectiniforme (Roem. et Schult.) H. L. Yang)

Fig. B2.3 Scattered seeds (Agropyron cristatum (L.) Gaertn. var. pectiniforme (Roem. et Schult.) H. L. Yang)

B3. Agropyron Mongolicum Keng

[Features]

Perennial sparse grass. Root-shaped, with sand cover and rhizome. Culms are sparsely erect, base nodes are often knee-curved, 90–100 cm tall. The leaves are 11–16 cm long, 2–4 mm wide. Spikelike inflorescence, erect, 12–14 cm long, 7–10 mm wide, with loose spikelets arranged obliquely upward. The caryopsis is oval, 4 mm long, light yellow. The root system is developed, with the characteristics of high seed setting rate, fast seed germination and high germination rate. The spring turns green early, the yellowing period is late, the green lasts for a long time. It is resistant to drought, wind and sand, barrenness, cold resistance, and strong adaptability.

[Habitat]

Born in dry grassland, sandy land.

[Use]

Forage use (Figs. B3.1, B3.2, B3.3 and B3.4).

Fig. B3.1 Seeds (Agropyron mongolicum Keng)

Fig. B3.2 Scattered seeds (Agropyron mongolicum Keng)

Fig. B3.3 Field and samples (Agropyron mongolicum Keng)

Fig. B3.4 Field (Agropyron mongolicum Keng)

B4. Agropyron Desertorum (Fisch.) Schult

[Features]

Perennial sparse grass. Rhizome, with sand sheath outside the fibrous root. The stalk stands upright, 25–63.5 cm tall. The leaves are 5–10 cm long, 1–3 mm wide, and are often rolled into a cone shape. Spikes are erect, cylindrical, 7–18 cm long, 25 mm wide; inter nodes between cobs 1–1.5 mm long; spikelets 5–10 mm long, 3–5 mm wide, with 4–7 small flowers; glume boat-shaped, ridges with sparse short hairs, the first glume 3–4 mm long, the second glume 4.5–5.5 mm long, the apex 1–2 mm long; lemma boat-shaped, 5.5–7 mm long, usually glabrous or sometimes on the back and sides. Veins have short spines to some extent, apex 1–1.5 mm long, palea ridge is sparsely covered with short cilia. Drought resistance and cold resistance are strong and lodging resistant.

[Habitat]

Most of them are found in dry grassland, sandy land, hilly land, hillside and lowland between dunes.

[Use]

Forage use (Figs. B4.1, B4.2 and B4.3).

Fig. B4.1 Scattered seeds (Agropyron desertorum (Fisch.) Schult.)

Fig. B4.2 Field and samples (Agropyron desertorum (Fisch.) Schult.)

Fig. B4.3 Field (Agropyron desertorum (Fisch.) Schult.)

C. Bromus L

C1. Bromus Inermis Leyss

[Features]

Perennial herb, rhizomes amphitropous. Stems 45–80 cm tall, upright, glabrous or glochis at lower part of the internode. Sheath glabrous and only isolated at the sheath mouth; ligule hard, 1–2 mm long, leaves lanceolate, 7–16 cm long and 5–8 mm wide, glabrous. The inflorescence is an open-branched panicle, 10–20 cm long, 3–5 branches of each internode. Each branch has 1–6 spikelets. Each spikelet is 12–25 mm long, containing 4–8 florets; spikelet has fine cilia; glume lanceolate, membranous margins; lemma broadly lanceolate, glabrous or slightly rough at the base, 8–11 mm long, 5–7 veins, often unawned or 1–2 mm long awned on the ridge; palea is shorter than lemma with hair on the ridge; anthers yellow, 3–4.5 mm long. Caryopsis as long as palea.

[Habitat]

Hillside grasslands, valleys, roadsides and riverside.

[Use]

Forage use (Figs. C1.1, C1.2, C1.3, C1.4, C1.5, C1.6 and C1.7).

Fig. C1.1 Scattered seeds (Bromus inermis leyss)

Fig. C1.2 Field and samples (Bromus inermis leyss)

Fig. C1.3 Plant 1 (Bromus inermis leyss)

Fig. C1.4 Plant 2 (Bromus inermis leyss)

Fig. C1.5 Leaves (Bromus inermis leyss)

Fig. C1.6 Spike (Bromus inermis leyss)

Fig. C1.7 Inflorescence (Bromus inermis leyss)

D. Stipa L

D1. Stipa Grandis P. Smirn

[Features]

Stalk 50–100 cm tall, 3–4 internodes, persistent withered sheath at the base. Sheath coarse or smooth when old, lower part usually longer than internodes; basal ligules 0.5–1 mm long, obtuse, ciliiform at the margin, stalks 3–10 mm long, lanceolate; leaf-blades vertically-rolled needle-like, puberulous on the top, glabrous underneath, basal leaves 50 cm long. The base of panicle wrapped up in the sheath, 20–50 cm long, branches thin, erect and scrolled; spikelets light green or purple; glumes 3–4.5 cm long, lanceolate, filiform at the apex, the first glume 3–4 veins, the second glume 5 veins; lemma 1.5–1.6 cm long, 5 veins, the joints a round of short-pilose at the apex, dorsal surface clothed with longitudinal short-pilose, basal disc acuminate with pubescence, about 4 mm long, awn geniculate, twisted, slightly coarse, column length of the first awn 7–10 cm and second awn 2–2.5 cm, awn needle curly, 11–18 cm long; palea as long as lemma, 2 veins; anther about 7 mm long. Flowering and fruiting May to August.

[Habitat]

Most of them grow on vast and flat wavy talllands.

[Use]

Grazing; forage use (Figs. D1.1, D1.2, D1.3 and D1.4).

Fig. D1.1 Seeds 1 (*Stipa grandis* P. Smirn.)

Fig. D1.2 Seeds 2 (*Stipa grandis* P. Smirn.)

Fig. D1.3 Plant (*Stipa grandis* P. Smirn.)

Fig. D1.4 Field (*Stipa grandis* P. Smirn.)

E. Leymus Hochst

E1. Leymus Chinensis(Trii)Tzvel

[Features]

Perennial, rhizome stretching or horizontal; fibre roots with rhizosheath. Stalk scattered, erect, 40–90 cm tall, 4–5 internodes. Sheath glabrous, residual sheath fibrous at the base, withered yellow; ligules truncate, carnassial at the top, papery, 0.5–1 mm long; leaf-blades 7–18 cm long, 3–6 mm wide, flat or involute, coarse at the top and margin, and smooth beneath. Spike erect, 7–15 cm long, 10–15 mm wide; spike-stalk tiny ciliiform on the margin, internode 6–10 mm long and 16 mm long at the base; spikelet 10–22 mm long, containing 5–10 florets, usually 2 born on 1 internode, or solitary at the apex and base, pink green, and turn yellow when mature; internode of spikelet glabrous, 1–1.5 mm long; glume coniform, 6–8 mm long, equal to or shorter than the first floret, not covering the base of the first lemma, hard with 3 veins, glabrous at the middle and lower part, coarse at the top, ciliiform at the margin; lemma lanceolate, the margin narrowly membranous, acuminate at the apex, dorsal surface has inconspicuous 5 veins, basal disc glabrous, the first lemma 8–9 mm long; palea is the same long as lemma, bifid at the apex, fine ciliiform or glabrous at upper half of the ridge and anthers 3–4 mm long. Flowering and fruiting June to August.

[Habitat]

Open plains, hills, floods, salt marsh and lowlands.

[Use]

Forage use (Figs. E1.1, E1.2, E1.3 and E1.4).

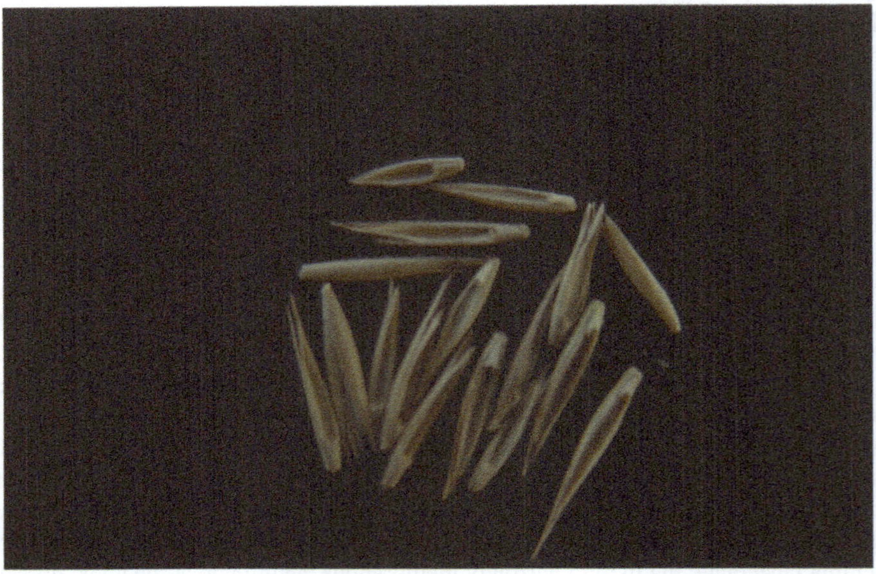

Fig. E1.1 Seeds (Leymus chinensis(Trii)Tzvel.)

Fig. E1.2 Plant (Leymus chinensis(Trii)Tzvel.)

Fig. E1.3 Spike 1 (Leymus chinensis(Trii)Tzvel.)

Fig. E1.4 Spike 2 (Leymus chinensis(Trii)Tzvel.)

Bibliography

Lu, S. L., Y. H. Sun, S. W. Liu, Y. C. Yang, Z. L. Wu, P. C. Kuo, S. L. Yang, C. P. Wang, and N. R. Tsui. Flora, reipublicae popularis sinicae, delectis florae reipublicae popularis sinicae agendae academiaesinicae edita. Tomus 9(3), Angiospermae Gramineae (3), Pooideae. Science Press. 1987.

The manufacturer's authorised representative in the EU is Springer Nature Customer Service Centre GmbH, Europaplatz 3, 69115 Heidelberg, Germany. If you have any concerns regarding our products, please contact ProductSafety@springernature.com

Printed and bound by CPI Group (UK) Ltd, Croydon, CR0 4YY
30/09/2026
02225307-0007